University building in 1809, and the organizational committee decided not to reconstruct it.

Although the University faced early struggles, there were significant achievements in science education. A charter for the Medical College of Ohio, ancestor of the present University of Cincinnati College of Medicine, was obtained by Dr. Daniel Drake in 1819. When he founded Ohio's Commercial Hospital and Lunatic Asylum, and affiliated it with the Medical College, he established the first teaching hospital in the Northwest Territory.

But long before that, in the 1820s, the academic department of the college, which offered courses in philosophy, rhetoric, Greek, Latin, and ancient history, folded. By the mid-1830s, a group of citizens revived Cincinnati College with medical and law departments, and later another academic department. Under respected educator Reverend William Holmes McGuffey as president, Cincinnati College for a time showed promise of becoming the flourishing city university its founders had intended.

Mitchel set educational precedents in astronomy as well as engineering. The charismatic professor was a genius in sharing the marvels of

ceiling and listening from above.

The Medical College of Ohio was an early archetype of university-hospital partnership. William Henry Harrison was instrumental in the State Legislature in passing an act to establish the college's affiliate hospital, also Cincinnati's first public infirmary.

PROLOGUE
1807-1858

The Means of Education Forever Encouraged

THE HISTORY OF EDUCATION in Cincinnati begins not long after the first settlers arrived in 1788. One year before that, the Northwest Ordinance provided for settlement and governing of the Northwest Territory and specified that "schools and the means of education shall be forever encouraged." From the beginning, those means in Cincinnati, especially in scientific and technical education, have fostered a tradition of innovation.

Cincinnati was incorporated as a town of nine hundred people in 1802, one year before Ohio joined

In his tireless efforts to establish the first American observatory, Professor Ormsby MacNight Mitchel personally visited twelve hundred citizens to sell three hundred shares of stock. In three weeks he recruited all of his shareholders.

astronomy with a lay audience. In 1842, when he began lecturing publicly to the people of Cincinnati, he used transparencies of drawings from European observatories to show his audience what astronomers were able to see through the world's greatest telescopes. The large, enthusiastic turnout for his lectures convinced him to build in Cincinnati the first observatory funded entirely by private citizens. Local residents formed the Cincinnati Astronomical Society that same year.

On November 9, 1843, John Quincy Adams laid the cornerstone at the dedication of the observatory. At age seventy-six he traveled to Cincinnati to give an address on the history of astronomy. In gratitude the society renamed Mt. Ida, the observatory site, Mt. Adams. The society members, who came from all economic levels, sent Mitchel to Europe to visit observatories and buy a telescope. After searching all of England and France, he finally found in Munich the lenses for a refractor telescope that, if bought by the society, would be unrivaled in the United States. To encourage the society to raise the required nine thousand dollars, Mitchel offered to direct the observatory gratis for ten years.

Triumph collided with disaster when the telescope reached America on January 23, 1845—four days after Cincinnati College burned to the ground. With his sole means of income gone, Mitchel took leave from the observatory the following October to give public lectures across the country. He continued to direct the observatory for fifteen years, while his lectures, published as books, attracted national popularity and acclaim.

Twenty-five years after its construction, the observatory was the setting for another technological achievement, when it became the world's first weather bureau. In 1869, director Cleveland Abbe began to collect weather information from across the country by telegram and issue storm predictions based on the reports. To finance the project, Abbe solicited funds from the Cincinnati Chamber of Commerce and obtained reduced rates from Western Union. Congress established the Federal Weather Bureau the following year. Abbe went to work for the new weather bureau in Washington in 1871. For his pioneering work, he was awarded medals by the National Academy of Sciences and the Royal Meteorological Society in Great Britain.

Meanwhile, other events were taking place which would bring higher education in Cincinnati out of

When the great refractor arrived from Germany, the Cincinnati Observatory joined the ranks of Greenwich, England, and the Imperial Russian Observatory at Polkowa, the world's two prestigious observatories at the time. This illustration of the refractor accompanied the word telescope in early editions of Noah Webster's dictionary.

EXCERPTS FROM THE MEMOIRS OF JOHN QUINCY ADAMS

The Cincinnati Enquirer
November 10, 1843
Cincinnati:
Friday Morning, November 10, 1843

☛The ceremony of laying the corner stone of the Observatory was gone through with yesterday in the midst of a continued and most cheerless rain. The Times, which, we suppose, is *au fait* in the matter, states the order of procession thus:

In front were the officers of the 1st Division Band; Cincinnati Light Dragoons; Cincinnati Guards; Light Artillery; Cincinnati Greys; Jefferson Riflemen; Washington Cadets; La Fayette Guards; Morgan Riflemen; Band; German Ridemen; Jackson Guards; Music; German Light Infantry; Montgomery Guards; Kosciusko; Board of Control and Committee of Arrangements; Mr. Adams and Suite; Astronomical Society; Band; Society of Book-binders, with their Banner; Band; Chamber of Commerce and Young Men's Mercantile Library Association; Mechanics' Institute.

The march to the hill, we apprehend, was anything but pleasant. Hundreds who would else have been eager spectators, remained at their homes in the city. The time occupied on the hill was short,—Mr. Adams making a brief address, and then the cortege returned.

Among other good things which our worthy Mayor said to Mr. Adams, in his welcome, was the declaration that "all yield admiration to your ardent love of liberty, and devoted attachment to the rights of man."

This saying of his Honor was *confirmed* by its adoption on the part of numerous colored gentmen at the classic ground of Bucktown, who displayed a banner yesterday, containing the inscription, "JOHN QUINCY ADAMS, the able defender of the rights of man."

Wednesday, November 8th. We breakfasted at Lebanon, and proceeded twenty-two miles to Cincinnati. The day was fine, and the sun shone bright. An immense crowd of people followed us to the Henry House, in Third Street, kept by Major Henry, where lodgings were provided for us. In front of this house there is a large balcony overlooking the street, upon which Mr. Spencer, the Mayor, delivered a complimentary address, welcomed me to the city, and introduced me to the assembled multitude, who answered by deafening shouts of applause. The arrangements for laying the cornerstone of the Observatory, and for the delivery of my oration to-morrow, were concerted with Professor Mitchel, Judge Burnet, and Mr. Greene; but the address on the spot, in the act of laying the cornerstone, was not yet prepared. Worn down with fatigue, anxiety, and shame, as I was, and with the oppression of a catarrhal load upon my lungs, I sat up till one in the morning, writing the address, which, from utter exhaustion, I left unfinished, and retired to a sleepless bed. I fear I am not duly grateful to Divine Providence for the blessing of these demonstrations of kindness and honor from my countrymen; but I dread still more the danger of being pampered and elated into vanity by them.

9th. I rose again at four, and before breakfast finished the address to be delivered at the laying of the corner-stone. A succession of visitors, invitations, and deputations ensued till ten o'clock, when a dense crowd of people gathered in the street fronting the Henry House, and a procession of the members of the Astronomical Society was formed. As we entered the carriage it began to rain, and we were obliged to raise the sides over our heads for shelter, and exclude the sight of me from the people, and of the people from me. The procession marched round sundry streets, the rain increasing till it poured down in torrents; yet the throng in the streets seemed not at all to diminish. It looked like a sea of mud. The summit of the hill was a circular plain, of which the corner-stone was the centre. At the circumference a stage was erected, from which my discourse was to have been delivered; but the whole plain was covered with an auditory of umbrellas, instead of faces. It was then determined that the discourse should be delivered to-morrow morning in the Wesleyan Methodist Chapel, and the corner-stone now laid—which ceremony I performed, and read the address which I had written last evening and this morning, to which the circle of hearers gathered around responded by three hearty cheers. My manuscript was so defaced by the rain that it was scarcely legible. The procession then dispersed, and the barouche returned and landed me at the Henry House, leaving the only part of the ceremony important to me still to be performed.

10th. At ten o'clock, the procession of the Astronomical Society went to the Wesleyan Methodist Chapel, the largest church in the city. I delivered about one-half of the oration which I had written, and was well received. The house was full and overflowing. I spoke nearly two hours, without a symptom of impatience or inattention of the auditory. Resolutions were adopted by the Astronomical Society, that the Observatory hill should henceforth be called "Mount Adams." I then shook hands with the members of the Astronomical Society, and all others in the church who desired it, and returned to the Henry House. We dined at Mr. John G. Pendleton's, with a party of about twenty—a splendid entertainment, in the highest style. Mr. Pendleton lives in a princely state. At seven in the evening we returned again to the Henry House, and thence . . . to the theatre, in the midst of the performance of the comedy of "London Assurance." We retired, and went to a ball at Mr. Springer's, an opulent citizen of Cincinnati. The ball was splendid, the banquet sumptuous and temperate, and the company genteel and lovely. Thus closes, blessed be God, one memorable day of my life.

To lay the cornerstone at the observatory, John Quincy Adams traveled for two weeks by train, steamboat, canal boat, and land coach from his home in Boston to Cincinnati. Despite his poor health, he stopped in many cities along the way to attend receptions and address the public. This last portrait of Adams hangs in the observatory, now affiliated with the University of Cincinnati.

Cleveland Abbe earned the nickname Old Probabilities among the clerks at Western Union. Abbe made his predictions based on reports from thirty volunteer observers across the country, who telegraphed weather information from as far west as Salt Lake City and as far south as Chattanooga.

dormancy forever. The will of Cincinnati businessman Charles McMicken, who died in 1858, made provisions for his homestead to be the site of a university. The bequest was written by McMicken himself:

"Having long cherished the desire to found an institution where white boys and girls might be taught not only a knowledge of their duties to the Creator and their fellow men, but also receive the benefit of a sound, thorough, and practical English education, and such as might fit them for the active duties of life; as well as instruction in all the higher branches of knowledge, except Denominational Theology, to the extent that the same are now or may hereafter be taught in any of the secular Colleges or Universities of the highest grade in the country, I feel grateful to God that through his kind Providence I have been sufficiently favored to gratify the wish of my heart." In that bequest he provided the means to build what is today's University of Cincinnati.

Learning How to Be a University

CHAPTER 1
1858-1902

Charles McMicken willed his homestead, on the hill where Vine Street meets Clifton Avenue, to be the site of a college. As heirs to the McMicken estate, the City of Cincinnati first spent twelve years fighting legal battles and resolving financial complications before establishing the University of Cincinnati.

BUILDING ANY UNIVERSITY from the ground up is a formidable and complex task, despite the visions of a generous benefactor. The deluge of complications in fulfilling Charles McMicken's vision were mostly financial, proving that his estate alone could not endow a college for boys and a college for girls. The McMicken family complicated matters by fighting to acquire legal status as the true heirs of the estate. But three years of court cases, one of which was eventually heard by the United States Supreme Court, only affirmed the City of Cincinnati's right to the property.

The coveted property turned out to be as much of a burden as a resource. Many buildings on the estate were dilapidated, and as existing rental property they did not generate much income. During the Civil War the high cost of living made it even more difficult for the city-appointed Board of Directors, who named their new institution McMicken University, to carry out the

THE SUPREME COURT RULED THAT BLOOD IS NOT THICKER THAN BEQUESTS

Although he was a slaveowner himself, Charles McMicken often gave money to the freed slaves he met in his travels between Cincinnati and Louisiana. With the president of Liberia, he planned and sponsored Ohio in Africa, a colony for free blacks in Ohio, Indiana, and Illinois wanting to settle in Liberia.

Although they were remembered in his will, Charles McMicken's relatives spent almost three years trying to convince the courts of Louisiana, and eventually the United States Supreme Court, that the entire estate should go to them, not toward building a university. Each niece and nephew (McMicken never married) was left either money or property. One unusual bequest went to nephew Charles McMicken Perin, who inherited one thousand dollars with a stipulation for an additional five thousand dollars if he would legally change his name from Charles McMicken Perin to Charles McMicken.

Franklin Perin, who married McMicken's niece and was the father of Charles McMicken Perin, first petitioned the Second District Court of Louisiana for the will to be declared null and void on the following grounds: By nature of being a city, Cincinnati was without the capacity to hold property in trust in another state. The property would be forever out of ownership if the City of Cincinnati became the perpetual owner. And the bequest had no devisee, since the boys and girls to be educated at the college did not exist at the time of McMicken's death, nor were they described with sufficient certainty.

Defendant Andrew McMicken, another nephew and also executor of the will, asserted the validity of the will, especially the section which left him and his family property in Illinois, Louisiana, Kentucky, and Texas; plots in the Spring Grove Cemetery; household furniture; and free occupancy in his uncle's house. The court ruled that Andrew McMicken's bequest was valid, but only the bequest of movable property, not real estate, to Cincinnati was valid. Both parties appealed the case to the Louisiana Supreme Court, which upheld the district court's decision on the real estate on the grounds that a foreign corporation was without capacity to take property. For similar reasons, the court declared Andrew McMicken's inheritance of out-of-state property null and void.

Meanwhile, relatives in New Orleans petitioned the United States Circuit Court to nullify the bequest of the McMicken homestead to Cincinnati on the grounds that a city is not capable of executing such a trust for want of corporate power. The defendants argued that by nature of being a city, Cincinnati had the right and obligation to possess the property in the interest of educating its citizens. And such an action was consistent with the general laws of the State of Ohio. The case on appeal, entitled Perin et al. vs. Cary et al., was heard by the United States Supreme Court. On February 25, 1861, Justice James M. Wayne rendered a decision which upheld Cincinnati's right to the estate and affirmed its ability to implement the bequest.

Adapted from "A Durable University at Last" in The University of Cincinnati: A Success Story in Urban Higher Education by Reginald C. McGrane terms of the will. New resources became available when the Cincinnati Board of Education passed a motion in 1868 to consolidate several high school funds for the cause of a university. The university board took immediate interest in the plan, and by 1870 the two boards had consolidated funds from Woodward and Hughes High Schools, the Ohio Mechanics' Institute, Farmers' College, Cincinnati College, and the McMicken fund to establish the University of Cincinnati. Two years later, the council authorized bonds for a building on the estate, and the board began to hire faculty. Salaries were attractive for the time at $3,000 per year for a full professor–almost twice the average of $1,750 at most eastern schools.

Woodward High School at Twelfth Street and Central Avenue was the last temporary university site when classes for freshmen commenced in October 1873 under borrowed Woodward faculty. The board hired four professors who began teaching the following September. Among themselves the faculty appointed Henry Turner Eddy, professor of mathematics, astronomy, and civil engineering, as dean of the university. Eddy was also in charge of the engineering program, which was one of three degrees offered. In its first twenty-five years, the university awarded about fifty engineering degrees.

In October 1875 classes moved to a new university building, still on the McMicken estate, at Vine Street and Clifton Avenue. That same year Edward Hyde, a Cornell graduate and professor of mathematics at Chester Military Academy, joined

The first university courses, open to freshmen only, were held in the afternoon at Woodward High School. Fifty students, out of one hundred fifty applicants, made up the first class. Forty of them were girls.

the faculty as assistant professor of mathematics and civil engineering. The program that evolved under Eddy and Hyde was the foundation of what would become the College of Engineering.

The University awarded its first degrees, a Bachelor of Arts and two Masters of Arts in 1877. In June 1878, at its first commencement ceremonies in Pike's Opera House, the University conferred seven degrees, including the degree of Civil Engineer to Thomas d'Aquino e Castro of Brazil. Another graduate, Albert Lilienthal, received the degree of Civil Engineer in 1878, but was not present at the commencement. Of the five engineering students who graduated the next year, three were also from Brazil. Then, for the next decade, the number of graduates in engineering averaged about four each year.

During its first twenty-five years the University of Cincinnati was in many ways learning how to be a university. Administrative attrition and financial uncertainty alternated with stability and growth. In 1877 the Board of Directors elected the Reverend Thomas Vickers rector of the university, dissolved Eddy's position as dean, and returned him to a regular faculty position. Changes under Rector Vickers included some additional technical courses. In 1879 Ward Baldwin, the first UC alum to join the engineering faculty, worked as an assistant in civil engineering for one year while earning a graduate degree. In 1881 Nelson W. Perry, a licensed instructor in metallurgy, was hired to teach metallurgy and assaying.

Rector Vickers was a controversial figure whose adversaries campaigned against his autocratic policies for most of his administration. Eventually, demands for his removal led to a formal board inquiry. The board officially confirmed its support for Vickers, but two years later, in 1884, he resigned. After Vickers' resignation, the directors again named Professor Eddy dean and rector pro tem while they looked for a president.

Edward Willis Hyde, left, and Henry Turner Eddy, right, were the first professors of mathematics and engineering at the University of Cincinnati. Eddy was also the first dean of the Academic Department and was acting president of the university for two years.

The turbulence of the Vickers era mirrored the atmosphere in Cincinnati in the 1880s. Bars and brothels flourished and crime was rampant. Rowdy events like prize fights, traveling circuses, and Buffalo Bill's Wild West Show were popular entertainments. Public services deteriorated in the crowded and growing city. Frustration with police and the courts erupted in the Courthouse Riot of 1884, in which fifty-eight people were killed. City politics were corrupted by political boss George Cox, who was elected to city council in 1877. Cincinnati earned the reputation as the worst-governed city in America while Boss Cox controlled everything from his table at Wielert's Saloon on Vine Street.

The work of another man named Cox restored confidence in the university. Jacob Cox (no relation to George), professor and dean of the Cincinnati College Law School, became president of the university on April 15, 1885. The former Civil War general and governor of Ohio brought stability and a new assessment of university operations and goals. Under President Cox's leadership the university's image improved, and close to 120 students were enrolled by the time he retired in June 1889. The proportion of engineering students increased to twenty percent of total enrollment.

Cox, who believed strongly in classical education, introduced more liberal arts courses into the engineering curriculum. By adding "philosophic studies" to the requirements and reducing technical courses as much as possible, he acted consistently with engineering education at other schools, which at the time was built on science, mathematics, languages, and social studies.

Professor Eddy, as dean of the academic faculty, succeeded Cox and again became executive head of the university. In May 1890 the Board of Directors named him acting president, but in September Eddy was offered the presidency of Rose Polytechnic Institute in Terre Haute, Indiana. The University of Cincinnati Board of Directors also offered him the presidency at a salary of $4,000—a substantial increase, though less than the $4,500 Rose had offered. Eddy accepted Rose's offer, and left Cincinnati after sixteen years of service to the university and the city. Ward Baldwin resigned from the University Board of Directors, on which he had served since 1884, and rejoined the faculty, replacing Eddy as professor of civil engineering.

After Eddy left, the board experimented with a different method of

The first University of Cincinnati building, in the upper left corner, was erected in 1875 behind the Christian Moerlein Brewing Company. By 1890 both expanding enrollment and industrial noise and fumes forced the university to relocate in Burnet Woods.

When it was completed, McMicken Hall provided the most spacious accommodations to date for every program. They included, from left to right, a lecture hall for chemistry, a biology laboratory, and a ladies' study room.

administration. Members of the faculty, in rotation, served as head of the university. Professor Hyde, for example, took a turn for a year in 1892, and again from 1897 to 1899. In all, during the eight years of rotation, four different faculty deans headed the university. This approach led to a number of problems, including lack of continuity, friction among faculty, power struggles between faculty and the board, and low morale. But because of financial constraints, the board believed that they could not afford to hire a full-time president.

Civil engineering also went through some administrative changes. In the early nineties, students entering the program earned a Bachelor of Science degree after four years of study. They could then choose to continue for a fifth year in an advanced program which included a professional thesis, and working in the "practical design of structures."

In 1892 the university renamed the Academic Department the McMicken College of Liberal Arts, in honor of Charles McMicken. The Academic Department included all programs outside of art and the observatory. The university catalogues listed civil engineering as a course of study or a programme of studies in the Academic Department. During this time, it was not uncommon for a catalogue to list departments, schools, and colleges as equals in hierarchy. This structure remained until 1900 when several new academic units, including the College of Engineering, were formally organized.

When President Cox retired, a plan to relocate the university was already underway, for the McMicken property was no longer suitable. The university building's steep hillside location was difficult to develop and had little room for expansion. The once-rural neighborhood had become an industrial district, filled with soot and fumes from nearby coal-burning boilers, heating plants, and breweries; the noisy Bellevue incline railroad to Clifton was not far away. Such an environment was hardly conducive to learning.

Hilltop areas became favorable sites once public transportation was available beyond the downtown basin. In 1893 the Cincinnati City Council issued $100,000 in bonds at four percent interest to build and equip a university building in Burnet Woods Park. The council had designated the area four years earlier, but intervention from the McMicken heirs, in one more attempt to claim the estate during this legal process, delayed the proposed move. Engineering faculty and students had roles in planning and implementing the relocation. Students in Ward Baldwin's class surveyed the site of the building and its steps up the hill from Clifton Avenue.

The cornerstone for McMicken Hall was laid September 22, 1894, and the university officially occupied the building the following fall. The new structure, facing Clifton Avenue, housed seminar rooms and classrooms, an assembly room, and a gymnasium. An adjacent annex for boilers, engines, and dynamos for heating and electricity made the building a self-contained unit. The first commencement ceremonies in McMicken Hall were on June 16, 1896, in the new auditorium.

Following McMicken Hall's completion, the university received the most significant gifts since its founding. Funding from local citizens between 1895 and 1899 totaled nearly a half million dollars. Much of the new construction directly benefitted the engineering program. Henry Hanna's gift of nearly seventy thousand dollars financed the construction and equipment of a new north wing of McMicken Hall, permanently housing the departments of chemistry and civil engineering. Hanna Hall, occupied in 1896, consisted of two connected structures, one of them 56 by 100 feet with two stories and a basement, the other nearly the same square footage with just one story and basement.

The original engineering section of Hanna Hall (about 9,000 square feet) included lecture, study, heliograph, and drafting rooms, as well as a museum. There were also hydraulic and testing laboratories for cement and metals. By 1900, the civil engineering department occupied the second floor, the basement, and part of the first floor of Hanna

The original McMicken Hall is on a list of prominent Cincinnati buildings designed by Samuel Hannaford which includes City Hall, Music Hall, St. George Catholic Church, and the City Workhouse. This building was razed fifty years after its construction and replaced by a new building containing 650,000 bricks from the original.

The Cincinnati Enquirer
September 23, 1894

A GREAT DAY

For the University—
Corner Stone of the New
Building Laid.

Despite the determination of the University Trustees to postpone formal ceremonies to the date for the opening of the new University building, there was a very considerable gathering in Burnet Woods yesterday to witness the laying of the corner stone of the new building.

The simple programme at first proposed, was elaborated as the day of the corner-stone laying approached until the exercises yesterday contained many features of great interest. There was first an invocation by Rev. David H. Moore and then brief, but happy addresses by Mayor Caldwell and Dr. Henry W. Bettmann, President of the Alumnal Association.

Dr. C. G. Comegys, President of the University Directors, then guided the corner stone to the place where it shall rest in future centuries, and delivered a most eloquent and fitting address. Dr. Benedict, Dean of the University, was to have been the next speaker, but illness prevented his presence at the ceremonies. The programme ended with an oration by General Samuel F. Hunt.

The following articles were deposited in the corner stone: A copy of the will of Charles McMicken, a copy of the will of Matthew H. Thoms, first catalogue of the university, last catalogue of the university, twenty-third annual reports of the Trustees, the Mayor's last annual message, the manual of the city, a copy of The Enquirer, a copy of the programme attending the corner stone laying, a pamphlet embodying the instructions to the architects of the building, a copy of the "Student's Annual," a copy of the "Cincinnatian," a copy of the McMicken Review and the manuscript of Judge Hunt's oration, entitled "Charles McMicken, Founder of the University of Cincinnati."

William Thoms Professors

The William Thoms Chair in Civil Engineering was endowed by Matthew Hueston Thoms, who never married and left his entire fortune to the University of Cincinnati. Since 1896, eight professors have held the William Thoms Chair.

1896 - 1900
Ward Baldwin, M.S.

1900 - 1902
Harry Thomas Cory, M.S.

1906 - 1929
Herman Schneider, Sc.D. (Hon.)

1929 - 1951
Howard B. Luther, Dipl. Ing.

1951 - 1958
Cornelius Wandmacher, D. Engrg. (Hon.)

1958 - 1979
Louis M. Laushey, Sc.D.

1979 - 1988
James McDonough, Ph.D.

1988 - Present
Paul Bishop, Ph.D.

Matthew H. Thoms

Hall, and the basement of the McMicken annex. By the next year, all three floors of Hanna Hall, plus all of the McMicken annex housed engineering programs.

One of the university's first endowments was to the civil engineering program. The William Thoms Chair in Civil Engineering was endowed in 1896 with $130,000 from Matthew Thoms, who named the gift in honor of his father, a pioneer Cincinnati builder. Ward Baldwin became the first William Thoms Professor in 1896. The instructorship in civil engineering was established as a regular position in 1893. It was filled by a graduate student working toward an M.S., or as of 1895, the five-year C.E. degree.

In 1898 Briggs S. Cunningham, a director of the university, announced a donation similar to Henry Hanna's for a south wing in McMicken Hall. Cunningham Hall, which would later be home to some of the growing engineering departments, was dedicated in June 1903, along with the Van Wormer Library and an athletic field.

By the end of the century, the university's financial situation improved enough to afford a full-time president. In July 1899 the board hired Dr. Howard Ayers, a graduate of Harvard University, Freiburg University in Germany, and head of the biology department at the University of Missouri. The board's mandate to "build up" the university was clear in the hiring agreement, which gave Ayers flexibility in appointing and dismissing faculty. Eight years of rotating faculty deans as the executive head had so lowered morale and robbed the faculty of their collective sense of purpose that the board agreed that wholesale removals might be necessary. In January 1900 Ayers requested resignations of several faculty in civil engineering, mathematics, physics, chemistry, and biology. Rumors that he had dismissed all faculty swept the university and the city, but in reality he had asked only eight out of 150 faculty to resign.

Some of the eight fought the request at first, but the directors supported Ayers and approved his recommendation. To help ensure the new president's success, past executives were the likely targets of removal, since the critics of faculty rotation had charged that each dean tried to make his position permanent, or to use it as a way to the presidency. In engineering, the board retired Ward Baldwin and replaced him with Harry Thomas Cory as the William Thoms Professor. They also retired mathematics professor E. W. Hyde and replaced him with Dr. Harris Hancock.

Local newspapers agitated on behalf of the professors, but as the changes bore fruit in increased attendance, improved scholarship, and acceptance by the students, the uproar eventually died. Ayers's initial turbulent years were followed by a period of comparative peace. Despite the controversy preceding their hiring, the new professors, who came to be known as the Old Guard, quickly gained respect and earned a reputation for quality and devotion to classical education.

Among Ayers's contributions were promoting graduate study and adding the Doctor of Philosophy degree to the curriculum. At his recommendation, the board made extensive improvements to the university buildings and grounds. They installed a new steam heating plant in place of the old hot air system, providing warm classrooms in the winter for the first time, and a water system to protect the buildings and supply the laboratories.

University Board member Briggs Cunningham (left) gave the university $60,000 for Cunningham Hall, named not for himself, but for his wife Susan. The surveying class of 1896 put a memorable smile on the face of Henry Hanna (right) when, during the dedication of Hanna Hall, they stood up and sang, "Hanna, Hanna, Hanna Hall, Henry Hanna gave it all!"

Within a year Ayers had organized an official College of Engineering, and named Professor Cory as dean. The 1900-01 University of Cincinnati catalogue, the first to name a College of Engineering, lists

FOR BACHELOR OF SCIENCE IN CIVIL ENGINEERING

COURSES.	Hours a Week. Sem. 1	Sem. 2	COURSES.	Hours a Week. Sem. 1	Sem. 2
Freshman Year.			**Sophomore Year.**		
Chemistry 1 and 2	5	3	Civil Eng. 2 and 3	4	4
Civil Engineering 1	--	1	English 2	1	1
English 1	1	2	Mathematics 4, 5	7	7
German 1*	4	4	Physics 2 and 3	4	4
Mathematics 1	4	4			
Physics 1	2	2			
Junior Year.			**Senior Year.**		
Civil Eng. 4, 5, 6 and 9	10	10	Civil Eng. 7, 8, 10	11	11
Mathematics 8	2	2	Mathematics 9	1	--
Physics 4	2	2	Electives	2	3

This course of study was required for the Bachelor in Civil Engineering degree in 1899. The course Civil Engineering I was also required for bachelor's degrees in chemistry, mathematics, astronomy, science, and physics.

The Cincinnati Enquirer
June 17, 1896

PLEASING
Commencement Exercises

In McMicken Hall of the Cincinnati University.

Eighteenth Graduation of Scholars From the Now Famous Institution—The Graduates.

The auditorium of the University Building last night was the scene of the eighteenth annual commencement of the University of Cincinnati. The immaculate white interior of the hall was in chaste and dignified keeping with the occasion. The background of the stage and overhead was a canopy of the United States flag, enveloping graduates, faculty and Board of Directors in an atmosphere of patriotism. Tall ferns and exotics on either side of the stage, with the class and university colors prominent, completed the hall decoration.

The programme, the consecutive numbers of which were announced by the Dean of the Faculty, Prof. Myers, began with a pleasing selection by the University Mandolin Club, under the direction of Prof. Joseph Donelly. He also directed the Glee Club, and it must be said to his credit that both organizations showed proper discipline, a good deal of energy and musical ability. A solemn, earnest and brief invocation was pronounced by Rev. Charles F. Goss, of Avondale.

Mr. William Charles Bauer delivered the first address, on the subject of "Science Work of the United States Government." It was a terse and intelligent presentation of this interesting subject. He examined in detail the history and the present condition of three distinct departments in this scientific work—the United States Fish Commission, the United States Coast Survey and the Smithsonian Institution. He explained the object of this scientific supervision and research to be first, the increase of knowledge by original investigation; secondly, its diffusion all over the world. The vigor, soundness and progressiveness of American science is acknowledged everywhere.

"SIGNIFICANCE OF WESTERN IRRIGATION."

Mr. Bauer was followed by Mr. Eldon Revare James in an address on the subject, "Significance of Western Irrigation." He traced the history of irrigating the arid Lands in the West from the Mormon immigration to Utah, in 1847, and spoke of the immense possibilities of this development of a new industry in the future. Irrigation attracted a superior class of people to the arid lands, for it put a premium on brains.

The University Glee Club sang "The Wild Bird Serenade," with a tenor solo by Mr. Morgan Stricklett.

Mr. Oscar Frederick Shepard, Jr., followed with a well digested, clearly presented address on the subject of "Two Main Problems in the Utilization of Energy."

Mr. Victor Emanuel Heintz, the speaker chosen by the class, delivered a masterly address on "A Conclusion From Imperfection." It showed a vein of solid terse reasoning, as well as oratorical talent.

Alice Lorena Leonard followed with the last address on the subject of "Our Inheritance From Greece." She showed how much of our modern culture, refinement and development in the fine arts we owe to the Greeks.

A selection by the glee club preceded the conferring of degrees by Major Frank J. Jones, Chairman of the Board of Directors. Mr. Jones prefaced his presentation with a few well-timed remarks addressed to the graduates and the members of the faculty.

THE GRADUATES

The graduating class and the degrees conferred were as follows:

Bachelor of Arts—

Allen L. Bramkamp.	Robert J. O'Donnell.
Frederick Cramer.	Simon Pelser.
A.J. Cunningham.	M.A. Rabenstein.
Stella M. Fortney.	Louise SpilmanWilson.
Alice L. Leonard	Wilson Ross Tenney.
Harry H. Mayer.	John M. Thomas, Jr.
Abraham J. Measing	

Bachelor of Letters—

Emma H. Baldwin.	M. A. McComas.
Sylvia Ewan.	Mary H. Moore.
Mary E. Griffith.	Leander D. Oliver.
Victor E. Heintz.	Frederick R. Peters.
Elizabeth H. Hill.	Edwin S. Smith.
Bessie F. Johnson.	Sarah V. Spillard.
Martha Jones.	Bryant Venable.
Ella Lytle.	Laura Whitson.

Bachelor of Science—

Charles W. Andrews.	Eldon R. James.
William C. Bauer.	George H. Kress.
Edwin Wm. Goode.	Dudley B. Miller.
Ralph Holterhoff.	Oscar F. Shepard.

Master of Arts—

Frederick Cohn, B. A. (University of Cincinnati, 1894.)

After the singing of the two numbers—"Lovely Rose," by Hermes, and "Roving," by Damrosch, by the University Glee Club, the benediction concluded the exercises.

After the commencement exercises a delightful reception was given in the parlors of the university building by the faculty to the Board of Directors and graduates. A magnificent oil painting of the late Dr. C. G. Comegys was put on exhibition for the first time, and attracted much attention. Refreshments were served and a delightful time enjoyed.

The alumnal reception will be held this evening.

faculty positions in three new specialties: mechanical engineering, electrical engineering, and technical chemistry.

At that point the university offered bachelor's degrees in four engineering fields. Three of them—civil, mechanical, and electrical— were in the College of Engineering; chemical engineering was still offered in the chemistry department. The year before the College of Engineering was organized, a chemical engineering program was offered in the civil engineering department. By the 1902-03 school year, a chemical engineering degree was among the college's offerings. Early in the development of chemical engineering, considerable debate questioned the need for the specialty to have an identity separate from chemistry. Lack of a major in chemical engineering at UC before 1900 may reflect the inconsistent leadership over the preceding years. By the time Ayers came to Cincinnati, mechanical and electrical engineering as professional specialties were well-established in the United States, and chemical engineering was also on the way.

In 1902, mechanical engineering professor Christian W. Marx began to organize evening classes in engineering. Marx came to Cincinnati from the University of Missouri at President Ayers's invitation. His classes were open to Cincinnatians who worked in the day and wanted to further their education at night. The classes, in roundtable format, were not intended to be college level, since most students did not have sufficient backgrounds in mathematics. The city funded a small part of the operating expenses, but Marx taught the classes without pay because there was no money available for his salary.

Evening courses lasted just one year because Marx could not encourage other faculty to donate their time. Renewed controversy surrounding President Ayers and his governance

This machine shop provided instruction to high school-age students when the university operated the Cincinnati Technical School for the city. After high schools established their own technical courses, the instruction on campus ceased and the College of Engineering acquired the school's building and all of its equipment.

President Howard Ayers alienated most of the faculty and students with his autocratic policies. But he had the consistent support of Law School Dean William Howard Taft who later, as Governor of Ohio, wrote a letter on behalf of Ayers to the Board of Directors when they began proceedings to dismiss him.

also hindered the program. Though he left UC in 1905, Marx continued to teach evening and Saturday classes in his downtown office.

During the Ayers years the university underscored its commitment to the community, while gaining the use of additional technical facilities through an agreement with the Cincinnati Technical School, a training program founded by private citizens fifteen years earlier. The university agreed to operate the technical school's program until the Cincinnati Board of Education could start a technical high school. In return, the technical school gave all its equipment to the university. Friends of the technical school financed a building on campus to house manual training, machine, and forge shops; a blacksmithing shop and foundry; biological, chemical, and physical laboratories; and a drawing room. By the 1902-03 academic year, three persons responsible for woodwork; forging and machine shop; and machine design and drawing were listed in the university catalogue as College of Engineering faculty.

The relationship with the technical school continued until 1907, when the city high schools introduced technical instruction to their curricula. The technical school went out of existence, and gave the building, which came to be known as Old Tech, to the College of Engineering for laboratory space.

The Ayers administration enjoyed an opportune time to develop the young College of Engineering. The United States was on the verge of explosive economic growth, fueled by a wave of scientific discovery and new technology. The need for professionally trained engineers would soon escalate at an unprecedented rate. At the University of Cincinnati, the college was positioned to embark on its challenge to educate engineers for the twentieth century.

CHAPTER 2
1902-1920

The Cincinnati Plan

At the time of this 1913 picture of the electrical engineering laboratory, regular five-year co-op students paid $75 per year for tuition. The lab fee was $2.50 per lab period per week. At their co-op jobs, students earned 10 to 19 cents an hour.

THE COLLEGE OF ENGINEERING was established amidst autocracy and controversy, in a university whose strength was in liberal arts. Yet only six years after becoming a college, in 1906, it was the site of a revolution in American engineering education. Although American universities in the 1900s were based principally on the English classical model, engineering was a classroom curriculum rather than an apprenticeship, as in Europe. Over one hundred colleges of engineering were established by the turn of the century; they were heavily rooted in theory and contained minimal practical content. Shop work in forging and operating lathes and presses had been part of technical schools' curricula for years, but it produced machinists and foremen, not necessarily engineers. As university laboratories in America gradually replaced shops, their work mostly demonstrated scientific theories and did not extend to the practical principles of industry. Employers often found that most new engineers could not effectively apply their skills until after they gained some practical work experience during a probationary period.

As a young civil engineering instructor at Lehigh University, Herman Schneider struggled with this theory/practice dilemma. The idea of cooperative education taking form in his mind had grown from his personal experience and philosophy. He noticed that students who best grasped the subject matter of his courses all had practical engineering experience. And the new graduates who gained work experience in college performed exceptionally well on their first jobs. This pattern reinforced Schneider's belief that his own practical experience in his teens and during college helped him understand course work more readily than his classmates.

He knew the importance of teaching fundamental theory, but did not understand why it was exclusive of practical application. So he explored incorporating part-time work for credit into the engineering curriculum, to make theory more understandable. Some practicing engineers agreed with the idea, but most believed that the challenge of

Herman Schneider began his professional career as an architect. From 1894 to 1897 he practiced in Cumberland, Maryland. For this portrait, he is joined by his wife Louise Bosworth Schneider.

preparing engineers for industry belonged to the universities. Most professors objected to his cooperative plan for fear that industrial exposure would turn their students into common laborers or boilermakers. Many in industry were cool to the idea of college boys entering the work force, boys who might feel superior to the full-time workers and not take their jobs seriously. Schneider, who had worked with miners, railroad workers, engineers, students, and businessmen, felt both views were unfounded.

By 1901 Schneider was looking outside Lehigh to start a new institution which could easily accommodate cooperative education. While he was making progress with some Pittsburgh corporations in creating such a college, other industrialists in that city announced their sponsorship of the Carnegie Technical Schools, which became the Carnegie Institute of Technology. These sponsors had no plans to use Schneider's idea, though Andrew

Carnegie later told University of Cincinnati President Charles Dabney that cooperative education would have saved him millions spent on shops in Pittsburgh.

After deciding that his plan would have a better chance in another part of the country, Schneider moved to the University of Wisconsin. One year later, Frank Dufour, a fellow Lehigh graduate who taught civil engineering at Cincinnati and wished to teach at Wisconsin, traded positions with Schneider. In the summer of 1903 Christian Marx met with Schneider in Philadelphia, and after a brief interview, recommended his appointment to the college as assistant professor of civil engineering. With an exceptionally diverse industrial community, Cincinnati appealed to Schneider as a potential environment for his plan. And the three-year-old College of Engineering offered several programs: technical chemistry, shop practice, and mechanical, electrical, and civil engineering.

The Pittsburgh disappointment was still fresh in his mind as Schneider schematically cultivated Cincinnati's interest in cooperative education.

Representatives from fifty-seven colleges and universities attended Charles Dabney's inauguration on Wednesday morning, November 16, 1904. Henry Turner Eddy, professor of mathematics and civil engineering at UC from 1874 to 1890, returned to campus as the University of Minnesota's representative.

President Charles Dabney was convinced of cooperative education from the start. Shortly after Dabney was inaugurated, Schneider presented his 1902 paper, "A Communication on Technical Education," to him. It impressed Dabney so that he immediately endorsed the plan, and it distinguished Schneider in the president's mind as a potential dean.

He met with industrial leaders, workers, craftsmen, and representatives from trade associations. Subtly and patiently, he convinced them that both education and industry would benefit if students were to spend half their time in school and half their time in industry. Meanwhile, at the university, administrative upheaval was resolved when the board replaced President Ayers with Charles William Dabney. The climate which followed that controversy was right for Schneider's idea.

President Dabney's background and vision turned out to be the perfect complement to the cooperative education concept. As professor of chemistry at the University of North Carolina, Dabney discovered valuable deposits of phosphate and tin ore in that state. He was director of the North Carolina Agricultural

Experiment Station, and served as assistant secretary of agriculture under President Grover Cleveland. As president of the University of Tennessee he worked to develop industrial education in the South. He was a strong supporter of the idea that municipal universities could fulfill the American ideal of democracy in education by enabling young men and women, regardless of status, to receive a university education locally. To uphold this vision, Dabney desired a closer relationship between the university and local industry, and cooperative education fit in naturally. Dabney was like Schneider in that he, too, experimented with the cooperative idea, and was therefore impressed with a paper Schneider sent to him in 1904, entitled "A Communication on Technical Education."

Schneider used that paper to compose a speech to the Cincinnati Society of Mechanical Engineers. By chance, a small newspaper article reporting the speech caught the attention of John M. Manley, secretary of the Cincinnati Metal Trades Association. Manley contacted Schneider to learn more about the plan, and became the first industrial representative thoroughly convinced of cooperative education. His enthusiastic support facilitated contacts with the first twelve firms to employ electrical and mechanical engineering students on a cooperative basis for one academic year. Two university directors, Frederick A. Geier, president of the Cincinnati Milling Machine Company (which became Cincinnati Milacron, Inc.), and Ernest DuBrul, president of Miller-DuBrul-Peters Company, were instrumental in convincing other companies and the university administration of the plan's merits.

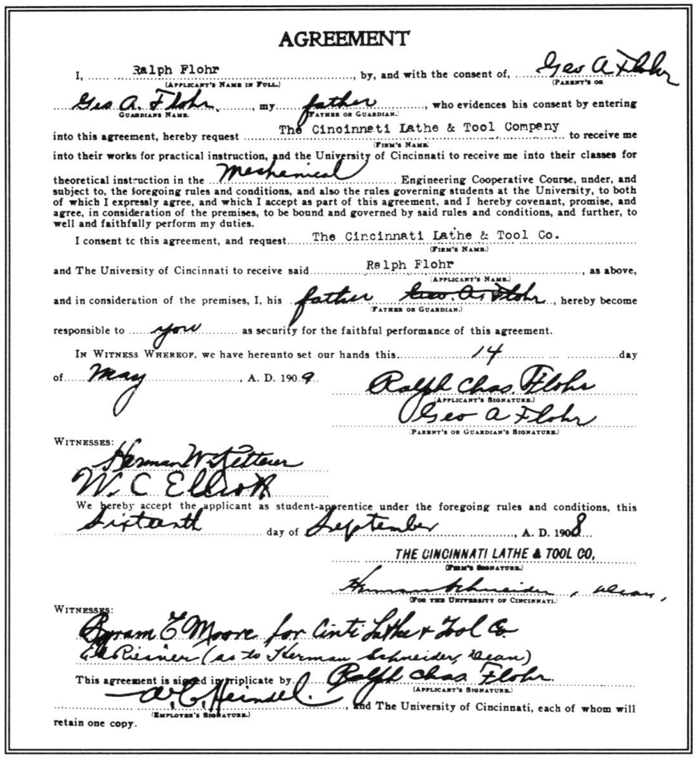

AGREEMENT

I, Ralph Flohr (Applicant's Name in Full), by, and with the consent of, Geo A Flohr (Parent's or Guardian's Name), my father (Father or Guardian), who evidences his consent by entering into this agreement, hereby request The Cincinnati Lathe & Tool Company (Firm's Name) to receive me into their works for practical instruction, and the University of Cincinnati to receive me into their classes for theoretical instruction in the Mechanical Engineering Cooperative Course, under, and subject to, the foregoing rules and conditions, and also the rules governing students at the University, to both of which I expressly agree, and which I accept as part of this agreement, and I hereby covenant, promise, and agree, in consideration of the premises, to be bound and governed by said rules and conditions, and further, to well and faithfully perform my duties.

I consent to this agreement, and request The Cincinnati Lathe & Tool Co. (Firm's Name) and The University of Cincinnati to receive said Ralph Flohr (Applicant's Name), as above, and in consideration of the premises, I, his father (Father or Guardian) Geo A Flohr, hereby become responsible to you as security for the faithful performance of this agreement.

In Witness Whereof, we have hereunto set our hands this 14 day of May, A. D. 1909.

Ralph Chas Flohr (Applicant's Signature)

Geo A Flohr (Parent's or Guardian's Signature)

Witnesses: Herman W. Fetten; W. C. Elliott

We hereby accept the applicant as student-apprentice under the foregoing rules and conditions, this Sixteenth day of September, A. D. 1908.

THE CINCINNATI LATHE & TOOL CO. (Firm's Signature)

Herman Schneider, Dean, (For the University of Cincinnati)

Witnesses: Byram E. Moore for Cinti Lathe & Tool Co.; ... (as to Herman Schneider Dean)

This agreement is signed in triplicate by Ralph Chas Flohr (Applicant's Signature), A. C. Heindel (Employer's Signature), and The University of Cincinnati, each of whom will retain one copy.

In the original cooperative agreement between student and employer, the student's father was responsible to the company for his son's "faithful performance of this agreement." In their first semester the co-ops earned ten cents an hour. By their last semester they earned twenty-one cents an hour.

Dabney saw that through cooperative education, the college would save the expense of constructing shops, and could therefore acquire its own building with lecture halls, study rooms, lounges, a library, and laboratories. He urged the Board of Directors to consent to the trial year. By a five-to-four vote they passed the plan, "for the failure of

THE FIRST CO-OP COMPANIES

In the 1906-1907 University of Cincinnati Record, thirty-one companies make up the first list of those who joined the university in cooperative education.

American Valve & Meter Co., Valves and Meters
M.L. Andrews & Co., Special Machinery
Bickford Tool & Drill Co., Drills
Bullock Electric Mfg. Co., Electrical Equipment
Cincinnati Electric Tool Co., Electric Tools
Cincinnati Machine Tool Co., Drills
Cincinnati Milling Machine Co., Milling Machines
Cincinnati Planer Co., Planers
Cincinnati Punch & Shear Co., Punches and Shears
J.H. Day Co., Bakers' Machinery, Special Machinery
Greaves-Klusman & Co., Lathes, Woodworking Machinery
Hilbert Machine Co., Special Machinery
Hisey-Wolf Machine Co., Electric Tools
Houston, Stanwood & Gamble, Engines
Lane & Bodley Co., Engines
R.K. LeBlond Machine Tool Co., Lathes
Lodge & Shipley Machine Tool Co., Lathes
Lunkenheimer Co., Valves
John H. McGowan Co., Pumps and Engines
Miller, DuBrul, & Peters Mfg. Co., Cigar and Cigarette Machinery
John B. Morris Foundry Co., Radial Drills and Special Machinery
Mueller Machine Tool Co., Radial Drills
National Machine Tool Co., Special Machinery
Oesterlein Machine Co., Milling Machines
Queen City Machine Tool Co., Shapers
John Steptoe Shaper Co., Shapers
U.S. Cast Iron Pipe & Foundry Co., Special Machinery
Watkins Laundry Machinery Co., Laundry Machinery
Weir Frog Co., Frogs
D.T. Williams Valve Co., Valves
Steel Foundry Co.

At their orientation in September 1906, the first co-op students were greeted with encouraging speeches by President Dabney, Dean Schneider, John Manley, and Ernest DuBrul. In reminding the students that they would face particular challenges in the unprecedented system, DuBrul suggested that if they had not already picked their class colors, they should consider black and blue.

which [they] would not assume responsibility." In 1906, Schneider became dean of the college and assumed the William Thoms Professorship, which had been unfilled for three years. With the appropriate authority he could finally develop cooperative education according to his own vision.

Twenty-seven students comprised the first co-op class in 1906. Twelve studied electrical engineering, twelve studied mechanical engineering, and three studied chemical engineering. The one hundred seven other students remained in the traditional full-time academic year. In the co-op course two students shared one job, alternating one week at school with one week on the job for six school years; it shifted to full-time work experience during the summer months, with two weeks vacation. As program coordinator, Schneider visited the students and employers at their jobs.

In 1909 civil engineering joined the co-op plan and brought with it the construction, concrete, and railroad industries. During that year Northeastern University in Boston became the second school in the country to offer cooperative education. Schneider continued to publicize the plan by addressing local and national industrial groups, publishing the results of his venture, and accepting invitations to initiate the program at other schools. He was also in demand as an arbitrator in labor/management disputes, and distinguished himself for resolving the dispute between the Cincinnati Traction Company and its union in the 1913 Cincinnati streetcar strike.

Eight hundred students applied to the co-op program in its second year; the third year it drew two thousand. Even with only half the co-op students in class at any one time, the physical limitations of the Old Tech building, which housed the College of Engineering, prevented any rapid increase in enrollment. President Dabney saw the college's growth as an inevitable catalyst for the university's growth. He joined the board, alumni, and industry representatives in a 1908 building campaign, which focused on co-op education's national recognition. City Council was well aware of the constant stream of visitors who came to the city to observe "the Cincinnati plan" in operation, and of magazines and journals that published articles about it. In 1909 they approved a bond issue of $450,000 to build a new engineering building and power plant.

Dabney put Schneider in charge of plans for the building. Schneider envisioned a quadrangle devoted to engineering, scientific research, business administration, and applied arts. He planned for gradual development, but presented his proposal so convincingly that he won immediate approval for the three-building quadrangle. The first building was ready for occupancy in September 1911. Cincinnati publisher Francis

Howard Baldwin left a $700,000 bequest in trust for the university in 1916, and in 1930 the building was named Baldwin Hall. The Chemistry Building was also completed in 1916. Swift Hall was completed in 1926 with the help of the Swift family, owners of the Eagle Picher Company.

By 1910 it was apparent that the six-year course was longer than necessary. The college extended the alternating weeks of school and work through the summer months, which provided the same classroom time in five years of eleven months each. This schedule was also more convenient for employers, since they did not have to provide summer work for twice the usual number of students.

As the student population increased, Schneider had to relinquish some coordination duties to department heads. In 1911 Professor Curtis Myers became the college's first professor of coordination, with Max Robinson as his assistant. After World War I the office grew into a Department of Coordination, which ensured and maintained successful matches between students and employers.

By 1912 the non-co-op student population dropped to forty. Several upperclassmen began alternating school and work every two-weeks, which was so successful that it became the official co-op schedule the following year. Fifty-five cooperative employers represented the machine tool, electrical, chemical, and utility industries in Cincinnati, Hamilton, Middletown, Dayton, Springfield, Columbus, and smaller surrounding towns. The engineering faculty and coordinators continuously monitored both the academic curricula and their related practice programs. Purely descriptive material was eventually removed from courses to accommodate a deeper study of fundamentals. First-year course work became uniform, so that if a student discovered early in his work experience that he was better suited for a different field, he could easily change to another engineering major.

On campus, engineers were filling leadership positions as they formed social groups within their majors. The Co-op Club was founded in 1908 to promote unity and fellowship among all engineering students. Between 1910 and 1919 the students established the Chemical and Metallurgical Engineers' Club, and University of Cincinnati branches of the American Society of Mechanical Engineers, the American Institute of Electrical Engineers, the American Society of Civil Engineers, and Tau Beta Pi, the national engineering fraternity.

The fine arts had always been important to Dean Schneider, so he introduced his students to art as both a hobby and a way of life. He believed his students could transcend the stereotype of an engineer as a narrow technologist if they

Baldwin Hall was constructed in 1911 by architectural firms Tietig & Lee and Garber & Woodward, at a cost of $300,000. In 1930 the building was named for Cincinnati publisher Francis Howard Baldwin, who left his entire estate to the university, with no conditions attached.

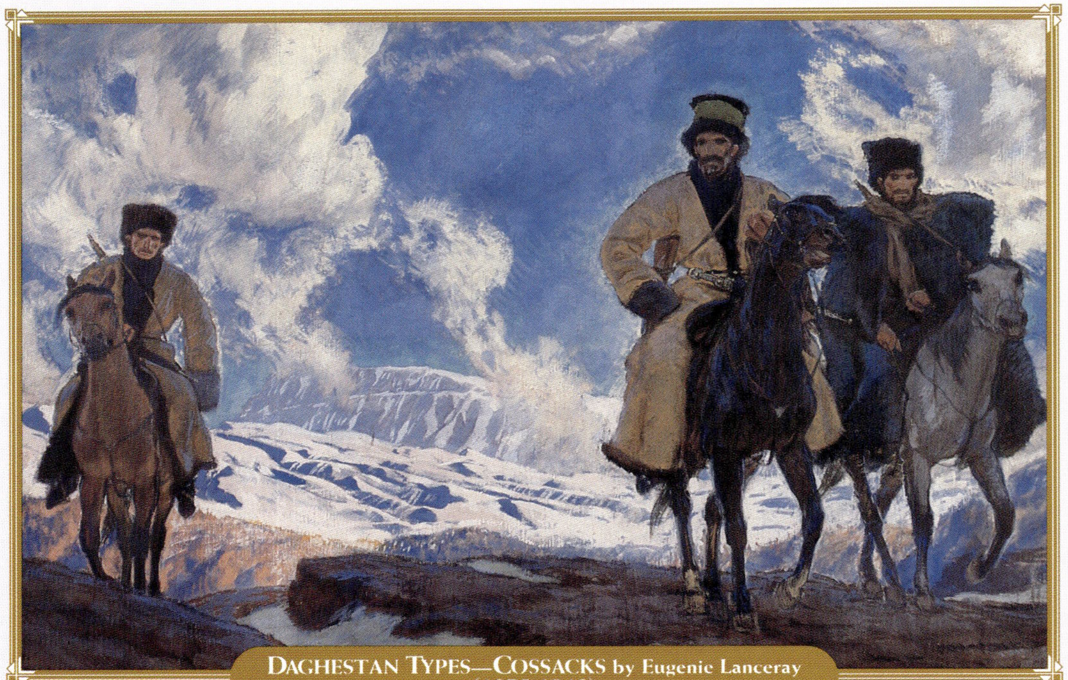

DAGHESTAN TYPES—COSSACKS by Eugenie Lanceray
(1875-1946)

THE ART of ENGINEERING

Herman Schneider wanted to make renaissance men and women of his students. He saw integrity in an education that could bring together skill in science and knowledge of art and literature. Such a background in college, he reasoned, would be the start of an education for life.

As part of his educational philosophy of joining the practical with the aesthetic, he regularly gave informal talks about art to his students. His interest in the art discussions stemmed partly from his own hobbies of painting and sculpture, but his primary reason behind the lectures was his belief in the distinct relationship between art and engineering. Every structure and design requires a balance and form to make it pleasing to the senses.

WINTER by Edwin A. Abbey
(1852-1911)

As he discussed art history, he spoke of art in southern Italy, of landscapes and bridges, of pottery and statuary. The students listened and joined the discussions. Then Schneider suggested they expand their interest by visiting art museums. The students replied that with their heavy academic and work

FOUNTAIN SQUARE IN THE RAIN by L.C. (Louis Charles) Vogt (1864-1939)

schedules, time was in very short supply. But having works of art around them on a daily basis would serve the same purpose.

The idea was a sound one to Schneider, who had long held a notion of a campus museum. The next best thing, a college art collection, satisfied his approach to educational democracy, where paintings and sculpture were available for everyone to study, regardless of social or economic rank.

The college began collecting art in 1916. For the most part, Schneider sat back and let his students carry out the plan, lending advice when they asked. After initially raising five hundred dollars, they commissioned Cincinnati artist Frances Wiley Faig to paint murals on the walls of the engineering library. In June 1917 Schneider unveiled the murals, which

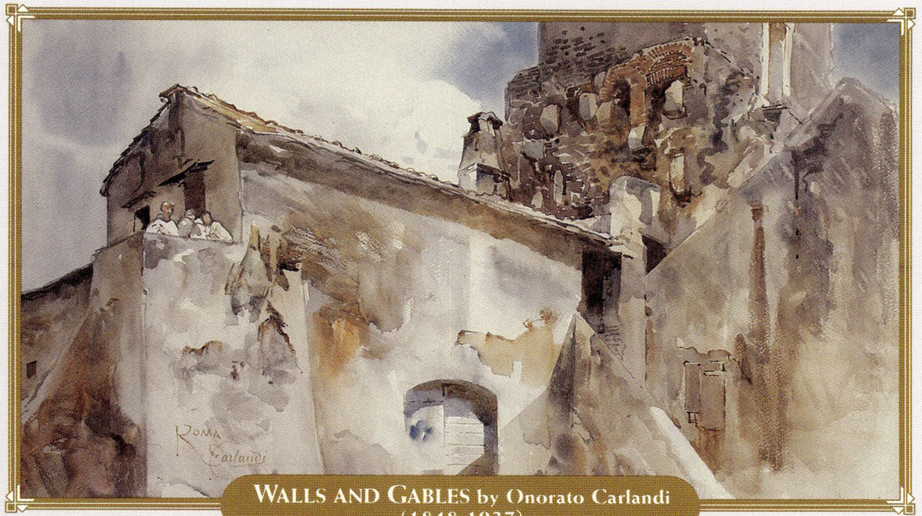

WALLS AND GABLES by Onorato Carlandi (1848-1937)

documented civic and industrial life in Cincinnati with the riverfront, woodlands, bridges, and buildings. At the dedication he said, "The engineer has learned that his creation must be a thing of beauty as well as utility."

The art collection rapidly grew over the years with gifts from alumni, students, and friends of the university. Mr. and Mrs. Charles Taft donated paintings by Degas and Rowlandson. Frank Duveneck and the family of Cincinnati artist Lewis Henry Meakin put their works on permanent loan. As diverse purchases were made of works by Martin Rico, Edwin Abbey, and Edward Henry Potthast, Schneider was asked by the students to advise them on acquiring paintings. His own tastes were broad, but he preferred works by Cincinnati landscape artists. As art graced the halls and alcoves throughout the decades, it became, as Herman Schneider wished, a daily experience for the co-ops.

Beginning in April 1917, students in the College of Liberal Arts and the College of Engineering participated in compulsory military drill through the Students' Army Training Corp. At the opening of the SATC and induction of men, October 1, 1918, President Dabney, at the edge of the platform to the left of the athletic field, addressed the troops.

could appreciate art and discuss kindred subjects intelligently. Although Schneider knew that some understanding came through course work and organized discussions, he believed real appreciation developed by close and frequent contact with the arts. To this end, he made the corridors of Baldwin Hall an art gallery of original paintings and sculpture, where he encouraged local artists to exhibit. Many students discovered that the exposure to art enhanced creativity in their engineering design projects. The class of 1916 began a tradition of leaving a monetary gift to the college to purchase artwork.

When the United States entered World War I, every phase of campus life, including the co-op program, changed. The College of Engineering suspended the requirement of industry experience, so that students could participate in the Students' Army Training Corps, a program implemented at over six hundred colleges and universities, which gave vocational training to Army draftees. At the University of Cincinnati, SATC was organized by the College of Engineering. A total of 1,772 men came to campus for two-month periods to train as auto mechanics or electricians. Thirteen temporary barracks spread over the entire campus. Students answered roll call, attended classes under strict supervision, and marched in formation between classes. Because of SATC, virtually every male student who had not already enlisted became a soldier.

Dean Schneider took leave of his university duties for the duration of the war to head the Industrial Services Section of the Ordnance Department in Washington. He became assistant to the chief of ordnance when the War Labor Policies Board was organized, and was a member of the Committee on Education and Training of the General Staff. During his absence, Professor Alexander M. Wilson of electrical engineering served as acting dean. SATC deactivated when the war ended in November of 1918, but the War Department appointed the University of Cincinnati to provide ongoing military training through the new Reserve Officers Training Corps. The College of Engineering required all physically fit male students to complete two years of military training.

When he returned from military leave, Dean Schneider resumed the ongoing job of assessing, improving, and refining cooperative education. Post-World War I reorganization within the college led to a merger with the College of Commerce, creating the College of Engineering and Commerce in 1919. The Department of Coordination was established in response to the increasing duties of the professor of coordination.

THE COLLEGE ON FIRE

The college art collection survived the fire of 1920 thanks to the students. As they evacuated the building, they grabbed the paintings from the walls and carried their treasures to safety in the cold January weather.

As the smell of burning rubber snaked through the upper floors of Baldwin Hall on a cold January day in 1920, the civil engineering co-ops assumed their electrical colleagues had simply burned out the motors on some of their laboratory equipment, as they had done before. But by noon smoke began to fill the hallways. On the roof it gathered in black billows and drifted to Burnet Woods. It wasn't the fault of the e.e.'s after all—Baldwin Hall was on fire!

Herman Schneider's dedicated and ever-efficient secretary, Anna Teasdale, sounded the alarm and led the evacuation. She shoved an armload of the dean's papers into a safe and then ushered the students toward the exits, making certain that on their way to safety they took the college's art collection from the walls. With Wessels and Duvenecks in arms, the co-ops marched single-file into the winter air of the quadrangle.

Firemen arrived on the scene and scrambled to lay hose from Clifton and University Avenues. Low pressure moved the water slowly, but once the hoses were in place the blaze was quickly extinguished. There was more than $11,000 in damage to the roof. The cause of the fire—in what was designed to be a fireproof building—was a burned out ventilation motor. Fortunately, there was little damage to classrooms or laboratory equipment.

A little over a decade later, the college was not so lucky. In February 1931, fire struck again, in the Basic Science Research Laboratory in Cunningham Hall, where inventor George Sperti was conducting research in food properties.

Again firefighters were hampered in their task, this time from the half-completed drive which ran in front of McMicken Hall. Sperti's team of researchers rescued what they could, carrying out as much equipment as they were able, while the firemen kept the blaze from reaching the adjoining McMicken and Hanna Halls.

The lab was almost totally destroyed, the roof reduced to rubble and ash, and supports and girders melted by the heat lay in twisted ruin throughout the area. The fire started in a section of the lab where experiments in light ray technology were conducted. Sperti and his colleagues were undaunted by the setback to their work. Within days they began design of a new Basic Science Laboratory and would continue their landmark efforts.

Assistant Dean George W. Burns became head of the new department, which was responsible for planning and developing co-op assignments to maximize their educational value to students. Coordination involved knowing each student's aptitudes and interests, as well as a familiarity with his studies and outside work. As co-op spread to other disciplines, the Department of Coordination expanded outside the College of Engineering to support other programs at the university.

Immediately after the College of Engineering and Commerce was established, a five-year program in commerce and business administration was developed by Schneider and Francis H. Bird, who headed the program. When the two colleges merged, existing evening courses in commerce initiated the Evening Division of the College of Engineering and Commerce. Designed like the co-op plan, it allowed adults who worked full-time to attend college classes. Dean Schneider appointed Professor Vincent H. Drufner, a UC graduate in civil engineering, to be supervisor of evening courses. The new college was supported enthusiastically by the faculty, who believed that knowledge of accounting and marketing was essential in the engineering profession, and also that students of commerce should have an understanding of technology and production.

As work opportunities outside Cincinnati became more available, some students were granted permission to expand their schedules to alternate four weeks of classes with four weeks of industry work. As a result, transportation was less of an obstacle, and increasing numbers of students who traveled long distances to study at Cincinnati were able to accept co-op jobs in their home towns. By 1920 all co-op students had assumed four-week schedules.

Also in 1920 President Dabney reached the age of automatic retirement. The Board of Directors received some requests to replace Dabney with William Howard Taft, a former dean of the Law School.

After two colleges merged to create the College of Engineering and Commerce, the faculty of 1919-20 posed for their first group photograph on the steps of Baldwin Hall. Fifth and sixth from the left are Dean Schneider and President Dabney.

Other faculty and administrators saw Schneider as the heir-apparent, but he firmly declined the position, preferring to remain dean of the College of Engineering. His lack of interest in the presidency was genuine, for he also turned down a candidacy for president at his alma mater, Lehigh University. In May 1920 the board elected President Frederick Charles Hicks from the old guard faculty recruited by President Ayers. He became UC's first professor of economics in 1900, dean of the College of Liberal Arts in 1907, and dean of the College of Commerce in 1912. When the commerce and engineering colleges merged, Hicks returned to teaching economics.

This group of 1918 co-op students called themselves The Gandy Dancers. From left to right are Joe Dave, Bob Sloan, Ray Hartung, Jerry Bevins, "Alabam'," Jack Harrod, and Norwood Geis. The College of Engineering and Commerce hired Geis to teach an evening course in economics when he was a senior. That appointment turned into a fifty-year career in teaching accounting at UC.

Other changes besides a new president greeted students when they returned to campus in the fall of 1920. Seven women were admitted to the College of Engineering and Commerce as co-op students. The University of Cincinnati had always admitted women, and by 1900 enough were enrolled to require a Dean of Women. But the idea of women in engineering did not attract serious consideration until Dean Schneider's secretary, Anna Teasdale, approached him on behalf of her friend Anna McFarlan. McFarlan, who was a member of the local Business Women's Club, had a niece who wanted to enroll in the college. Dean Schneider saw opportunities for educated women in industry, and granted permission for Ruth McFarlan and other qualified women to apply to the chemical and commercial engineering programs. He appointed Marjorie Stewart Palmer as their coordinator. The Business Women's Club provided support by selling the idea of cooperative study for women to employers. The first seven co-ed

ONE CO-EP'S JOURNEY

In 1925 Helen Norris Moore (Ch.E.), Kathryn Gillis (Com.E.), and Ruth McFarlan (Com.E.) were the first women to graduate from the College of Engineering and Commerce. "One Co-Ep's Journey," by Helen Norris Moore, appeared in The Co-operative Engineer in January 1934.

That first little group of seven co-eps was possibly a stepchild as far as the men students were concerned, but most certainly not as far as the Dean was concerned. I think one of the biggest thrills of our freshman year was the night that Dean Schneider, in spite of ill health at the time, took an evening off to meet with seven insignificant freshman, for the purpose of organizing a Co-ep Club. He regaled us with stories of the first class of Co-ops, told us that he expected great things from us and the classes of the future and made us feel as important as seniors.

Sometimes I wonder if we have even partly fulfilled his hopes for us. There were only three of that original seven who finished, two in the commercial and one in the chemical course. However, that first class laid the foundation for later classes, which really can achieve the high goal originally desired.

Such thoughts as these ran through our heads: "What good are all these reports?" "When will we ever be finished?" "How long will this job last?" "Will I ever get a raise?" "I wish we had an afternoon off to go to a show like the liberal arts people," etc., etc.—and then suddenly we were seniors. Everybody was nice and told us how glad they were that we had stuck it out. Well, I have always felt that probably Dean Schneider was more pleased than anybody. To show his apparent appreciation he gave us each a co-op pin, his own personal graduation gift to the first three senior women.

Then we were graduates looking for the ever elusive job. Even in 1925 jobs were not so plentiful that we could pick and choose between several which had been offered. In the first place women were not so desirable in chemical industries, and the second place, I had tacked a Mrs. to my name. Somehow or other, those two conditions made things rather difficult. We are all becoming more used to women working, even the married ones; but even so, it will never be as easy for a woman, even having the identical training as a man, to get a job. Occasionally the employer will take a chance, if he is favorably impressed and the woman agrees to accept a few dollars a week less. A lucky chemical [engineer] may get a job doing routine analysis in the laboratory of some chemical plant. The duties of answering the telephone, caring for the filing system for the laboratory, and even washing luncheon dishes are very apt to be added. But this is experience and permits watching for another job one step higher. I was acquiring my experience in this manner, when just as I was fed up with doing magnesium determinations, and anyway had been reading some biochemistry on the side, somehow fate stepped in with the chance to do graduate work. Fine! I started to school again, but what a difference! I had started in a new field—biochemistry and bacteriology, and there was still much to be learned about things of which I didn't even dream when I started.

Then came the best offer yet—an instructorship in my department. I found that I liked it all; that I never seemed to get satiated; that I was not skull-dragging, but actually doing constructive work; that each time I lectured I changed it just a little bit to make it more understandable; and so on.

In answer to queries of possible application to other college graduates, I should say it only indicates that every graduate must work out his own particular salvation or career—if one wishes to call it that. In order to advance and keep alert it is necessary to keep studying. Brains are like muscles, unless they are used, they tend to become useless.

co-ops quickly became known as the co-eps. Like the first co-op students, the co-eps were neither readily accepted nor easily discouraged. Many male students were not entirely pleased to attend classes which were formerly open to men only. Since their schedule followed the co-op calendar, it was hard for the co-eps to participate in extracurricular activities. But like the first male co-ops, they had Dean Schneider's support. He wanted them to succeed, and assisted them in starting their own organizations, one of which was Pi Chi Epsilon, the women's honorary professional society in engineering.

Although they were enrolled in the cooperative track as soon as they were admitted to the college, the first women took cultural courses in lieu of the most highly technical courses and could earn either a commercial or chemical engineering degree.
Standing, from left to right, are Ruby Schoen, Charlotte Atherton, Ruth McFarlan, Margaret Maynard. Seated, from left to right, are Kathryn Gillis, Helen Norris, Myrtle Hay.

The flourishing extracurricular activities included Hobby Hour, a concept introduced by Schneider to offset the intense six-day school week. Classes, with no electives, were in session from 8:30 a.m. to 5:00 p.m. on weekdays, and until noon on Saturdays. The dean had always systematically scheduled his own recreation, and believed that it increased his work efficiency. In organizing Hobby Hour, he encouraged students to do the same. Hobby Hour was scheduled in the middle of the day for students to pursue activities that were a complete change from work. Some played chess, pitched horseshoes, boxed, or fenced. Others read books, or took creative writing courses. Some students formed an orchestra, while a journalism group under English professor Clyde W. Park published *The Co-operative Engineer,* a magazine of technical articles, creative writing, and alumni news. Dean Schneider wrote a regular column for the magazine. Sketching, conversational French, rifle practice, and rhythmic dancing were popular pastimes for the co-eps.

The most remarkable milestone of 1920 was not any new program, but the end of an old one: the four-year, classroom-only program in engineering. In ceasing that option, the College of Engineering and Commerce became the first in the nation to graduate engineers with mandatory co-operative work experience.

The radical new program, which the university board had approved so skeptically that they disclaimed responsibility for its failure, made the pure classroom method of engineering instruction at UC obsolete.

In a 1905 address to the Cincinnati Society of Mechanical Engineers, Herman Schneider discounted the system of ranking students by examination grades and eliminating low achievers. In his words, that did not leave a "residue of the best material obtainable." He argued instead that through cooperative education a college could graduate the most qualified engineers. In this 1920 photograph he sits in the Baldwin Hall dean's office, next to the faculty room.

CHAPTER 3
1920-1939

The President's Desk in the Dean's Office

In 1921 the College of Engineering and Commerce occupied space in four campus buildings: Baldwin Hall, the Chemistry Building, Cunningham Hall, and the Old Tech Building. Clockwise, starting with Baldwin Hall in the back are Beecher Hall, the Van Wormer Library (later the Administration Building), the original Cunningham, McMicken and Hanna Halls, and the Chemistry Building. In the center, obstructed from view, is Old Tech. Swift Hall completed the quadrangle in 1925. To the far right is the gymnasium (later Schmidlapp Hall, the ROTC building). This photo was taken from the tower of Hughes High School.

THE TWO YEARS FOLLOWING World War I brought three landmark events to the college. It became the College of Engineering and Commerce, the first women students enrolled, and the cooperative program became mandatory for engineering degrees. Although they changed the course of the college forever, those events were merely the first milestones of the 1920s.

Early in the twenties, many chemical engineering programs across the country shared curricula with chemistry departments. Dean Schneider's desire to relocate chemical engineering, which was a subdivision of the Department of Chemistry, to the College of Engineering was representative of national trends, as was the chemistry department's resistance. They maintained that chemistry as an industry did not need technicians, and graduates in engineering belonged to that category. But the field of chemical engineering was changing, partly from an accreditation system initiated by

$$= \frac{a M_1 r_1 w^2 (\sin C \cos B + \sin B \cos C) + b M_2 r_2 w^2 \sin B}{I}$$

THE STUDENTS' FIFTY-YEAR HOBBY

For fifty years The Co-operative Engineer chronicled the academic, social, cultural, and professional lives of the students and faculty in the College of Engineering.

In October 1921, the first issue of The Co-operative Engineer, written by engineering students who dabbled in journalism during their Hobby Hour, was published. The co-ops intended their magazine to be more than a hobby, however. The quarterly featured practical articles on engineering and advice on building careers in their professional fields. It also provided a link back to campus for their fellow students engaged in the work phase of their co-op education. Regular columns featured campus events, Bearcat sports, alumni notes, and profiles of UC administrators. The first issue included a long article on Frederick Hicks, the new university president.

That inaugural issue of thirty-six pages also contained discussions of air mail technology and changes in airplane engine construction since the Great War. "Moments of Inertia," a regular page of jokes and cartoons, premiered in the first issue. Under first editor Harry Pockras, C.E. '21, the magazine staff declared that though many of the articles would have a technical bent, they also welcomed submissions of short fiction, poetry, and essays on the arts from students, faculty, and alumni. Though the yearbook, Cincinnatian, usually contained similar work, The Co-operative Engineer was the first campus literary magazine since the demise of the McMicken Review at the turn of the century.

Dean Schneider also used the magazine as a forum for his ideas on engineering education. In his regular column he encouraged students to become well-rounded citizens and professionals. In one early issue he offered advice on giving speeches: "A plain talk on a thing one knows, told as one would tell it to a man on the curb, is a good speech." The gist of his advice was to keep speeches simple and conversational. The editors noted at the end of the dean's essay that his writing practiced what it preached—it was composed entirely of one-syllable words! The popularity of The Co-operative Engineer quickly grew.

Advertisers like General Electric, Crosley Radios, Procter & Gamble, Union Gas & Electric, and Cincinnati Milling Machine Company, were typically employers of the co-ops. The range of articles expanded as the first years passed. Readers were given advice on marketing and advertising their designs; co-ops reported on their work experiences; co-eps voiced their own concerns as women engineers. Current engineering projects in Cincinnati, throughout the United States, and around the world were analyzed and debated; the Panama Canal, iron ore in the Great Lakes region, recycling wood waste into energy, river trade on the Ohio and the renovation of Cincinnati's Music Hall were all given attention in The Co-operative Engineer.

In the summer of 1943, the magazine suspended publication "for the duration," as Robert Gowdy, who was then dean, said in his final column. Like most campus activities, it was overshadowed by wartime priorities. "The duration" lasted until the spring of 1946, when The Dean's Page and the rest of the magazine returned. The Co-operative Engineer finally ceased publication after 1975. Over the course of its existence, editorial changes and new features continuously evolved. Some became more technical and others, like the Engine Charlie calendar and campus calendar girl, were arguably not. But the humor, alumni activities, and literary pieces thrived.

The Co-operative Engineer is a valuable legacy of those fifty years. Schneider's columns show him to be not only an innovative educator, but also a visionary—a person who accurately foresaw professional and social developments in the world. The thousands of pages give a clear view of engineering education past, present, and future. And in the frequent articles which detailed campus building projects like Nippert Stadium, Memorial Dormitory, Wilson Auditorium, the various Rookwood fountains, and many others, The Co-operative Engineer helped to document the university's history.

$$= \frac{a M_1 r_1 w^2 (\sin C \cos B + \sin B \cos C) + b M_2 r_2 w^2 \sin B}{I}$$

"Tuition was $200 a year. Freshman co-op jobs earned 35 cents an hour while senior co-ops earned $125 a month. Room and board (two meals a day) was $8 a week. Street car tickets were four for 25 cents. Least problem on campus: automobile parking!"
—Ransom D. Landon, C.E., M.S. '27

the American Institute of Chemical Engineers. So the university board granted Schneider the authority in 1921 to hire Reuben S. Tour as the first head of the Department of Chemical Engineering. Tour was educated at the University of Michigan and was a chemical engineer for the US Ordnance Department before coming to the college. This new department offered degrees in chemical engineering and metallurgical engineering.

The vision for a Department of Architecture in the college emerged from Herman Schneider's own work as an architect and a construction engineer. He saw architecture as the core discipline in a cluster of programs in applied arts, to join the artistic side of architecture with the pragmatic side of civil engineering. That department was established in 1922, with Professor Cecil Baker as director.

Growth continued in 1923 with the six-year cooperative program in general engineering, which led to dual degrees: a bachelor of engineering and master of science. The following year, the college added geological engineering to the co-op program, when the Department of Chemical Engineering discontinued the metallurgical engineering degree, and offered instead two new options in industrial chemistry and metallurgy.

In 1924 the college also began offering courses in engineering through its own evening division, separate from the Evening College. The program grew to include courses in applied arts by 1926. Under Professor Vincent H. Drufner from the College of Commerce, enrollment in the evening division grew from four hundred in 1920, to more than three thousand by 1928.

In 1925 the rapidly-growing Department of Architecture became the School of Applied Arts, with Dean Schneider as director. A year later Swift Hall, which completed the engineering quadrangle, was

In 1925 the School of Applied Arts was granted studio space in a room of the boiler house, where dynamos once generated electricity for the campus. The constant precipitation of soot forced students to cover their clothes and work, and made the lighting dim and hazy.

"As part of my thesis, I was asked by Professor Alexander Wilson to design a lighting system that would make night football games feasible for Carson Field. The task included the structural changes required to accommodate additional floodlights. Four G.E. thousand-watt incandescent floodlamps with a twenty-degree beam angle were installed on each of the existing wood poles—four on either side of the field. They were supported on two pairs of six-foot crossarms. The original five hundred-watt Westinghouse type L-11 floodlights, one at the top of each pole, were cleaned and remounted. Transformer capacity, then in the gymnasium building, was increased and all related wiring was replaced with conductors of appropriate capacity. The new lighting system was energized some time after the middle of June 1923. I recall climbing each pole—the poles had steps—to adjust the aim of the lights so that illumination was uniformly distributed. And, of course, I saw the first football game played under electric lights at UC in September of that year. The University News took note of the event with a complimentary editorial; its laconic but flattering heading stated, 'It Works!' "
—Jack B. Silverman, E.E. '23

"My first co-op job was with the Union Gas and Electric Company (UG&E). I was on the 'Cable Gang' with three other co-ops. Our job consisted of installing underground electrical cables under the streets of Cincinnati, changing over the distribution from direct current to alternating current. Each pull was a challenge because we had to invent a new rigging for each manhole, so that the cable under the street could be pulled by the winch on our truck. The whole system was very much overloaded. Finally, on a rush hour just before Christmas, the whole system let go. A cable burned out between manholes. Downtown Cincinnati was dark. Elevators got stuck between floors, meat thawed out in the freezers, and the traffic jam was unimaginable. We all got in plenty of overtime pulling in new cables."
—William J. Wenzel, C.E. '32

The Tanners' Council of America was the first organization to join the college's Institute for Scientific Research. The laboratory which it funded at UC was only the second endeavor in that industry's history to research the science of leather tanning.

finished in time to provide space for the new school, and for the Department of Electrical Engineering, which had outgrown its space in Baldwin Hall. By 1929 the cooperative option in applied arts attracted students in ceramics, landscape design, interior design, and art in industry.

The most comprehensive program in the School of Applied Arts was art in industry, which studied art in the design of everyday objects. Through cooperative relationships with industry, students had opportunities to design perfume bottles, jewelry, refrigerators, playing cards, book jackets, furniture, table lamps, and textiles. The curriculum also extended beyond functional consumer goods to package design and advertisement layouts.

Working partnerships between the University of Cincinnati and industry extended beyond engineering to other sciences at the Institute of Scientific Research. The institute expanded the concept of cooperative education by joining the university with industry at the research level. Schneider's plan was to build research laboratories, sponsored by industry, on campus. Staffed by university scientists, each laboratory would be devoted to research for its sponsoring industry.

Like cooperative education, the new idea met resistance at first. One common criticism, which

"I took economics in 1929-30. What was taught during the first semester before the stock market crash and what was taught after the crash offered interesting contrasts."
—**Robert E. Watkins, Com.E. '31**

ARCHITECTURE'S MODEL DOME

Is it possible for a basilica dome to simply disappear from campus? Although it wasn't a full-size structure crowning a real church, the College of Engineering possessed a scaled replica of the most famous one, St. Peter's Basilica. And they lost track of all twenty feet of it.

The story of the St. Peter's dome began in 1922 when Herman Schneider established the Department of Architecture with a five-year co-op program. Shortly after, he wrote to Gorham Stevens, director of the American Academy in Rome, asking for architectural models for the new students. Stevens was enthusiastic about the idea and immediately set his own students, all of whom were honors graduates of American architectural schools, to construct a model of the St. Peter's Basilica dome.

Michelangelo himself constructed a twenty-foot model of his dome design in 1588. Dogged by interfering dignitaries and labor problems when building the basilica, the artist feared St. Peter's would not be finished in his lifetime. So he constructed a wooden model to show the exact method of completing the dome. By doing so, he insured his plan against alterations by other artists who wished to replace his designs with theirs.

Stevens' plan was to reproduce that model. The Vatican granted him permission to build the replica, and stipulated that no other replica would ever be commissioned. The students in Rome built the model for their own education, so the only cost incurred by UC was packing and shipping. It was twenty feet high, twelve and a half feet in diameter, and large enough for students to walk inside and view the reproductions of the paintings on the inside of the actual dome.

When it was finished in 1925, the dome arrived in New York City and was exhibited there before it was shipped to Cincinnati. Upon its arrival, the Cincinnati Art Academy displayed it before sending it to the College of Engineering library. It was finally housed in Swift Hall, where students in the School of Applied Arts could study it. When it was exhibited to the public for the thirtieth anniversary of cooperative education in 1936, six thousand people walked through it.

Then it seemed to disappear. By World War II it was used occasionally for study, but for most of the time it sat in storage. When questions were asked of its whereabouts every few years, no one knew what happened to it. The dome was eventually found in a room beneath Nippert Stadium, but had fallen into disrepair and was then torn down.

"In those days UC was known as a streetcar university, since the students came largely by that means from all parts of the city. Riding the Crosstown was a social occasion since one knew almost everyone aboard by name. Those were small cars with four wheels and were usually packed with students. One day someone discovered that if all those on the back platform momentarily took the weight off their feet and then lowered it quickly, the car could be rocked on its four wheels, much to the chagrin of the crew. The Hughes High School students who came along on a slightly different schedule had not been exposed to elementary physics and were content with kids' stuff, like picking a trolley off the overhead wire now and then."
—P. Willard Crane, Ch.E. '26

came from industry, was that drawing on knowledge from just one university would cut an industry's competitive edge. But for several years the college had conducted research in basic sciences to address specific industrial questions from the shops that employed co-op students. Some upperclassmen used the problems as senior thesis topics, and Schneider was determined to build on this activity. The outcome was The Institute of Scientific Research, comprised of three laboratories: the Leather Research Laboratory of the Tanners' Council of America, the Research Laboratory of the Lithographic Technical Foundation, and the Basic Science Research Laboratory.

The institute's first member was the Tanners' Council, who in 1920 chose the College of Engineering and Commerce as the site of their research laboratory. The college and the council formally established the Institute of Scientific Research as the parent organization for the Tanners' Laboratory that same year.

The lithography industry formed the Lithographic Technical Foundation for the purpose of joining the Institute of Scientific Research in 1925. The institute extended the cooperative relationship to trade associations by providing university scientists and engineers to research for a particular industry, which in turn sponsored the laboratory.

The Tanners' Laboratory investigated problems in chemical science for the benefit of the tanning industry specifically, and for pure science in general. Dr. George D. McLaughlin directed the laboratory. The Tanners' Council funded their own research building, which was completed and presented to the university in 1924. At the dedication, Tanners' Council President Fraser Moffat shared the council's reasons for choosing the University of Cincinnati to operate the laboratory: "The committee, after visiting various colleges in the United States, selected the University of Cincinnati as the college having the broadest vision and the surest grasp of the relation between science and industry."

The printing industry joined the Institute of Scientific Research in 1924, when it sponsored a labora-

"While I was at UC, daily quizzes were frequent but unpredictable. I conceived the idea of running a pool on whether a quiz would be forthcoming in each class. Imagine the consternation of the professor when, upon announcing a quiz, one student would shout, 'Whoopee!' He had won the weekly jackpot."
—George W. Gasper, Ch.E. '29

BEFORE PRINTOUTS AND STUDENT NUMBERS

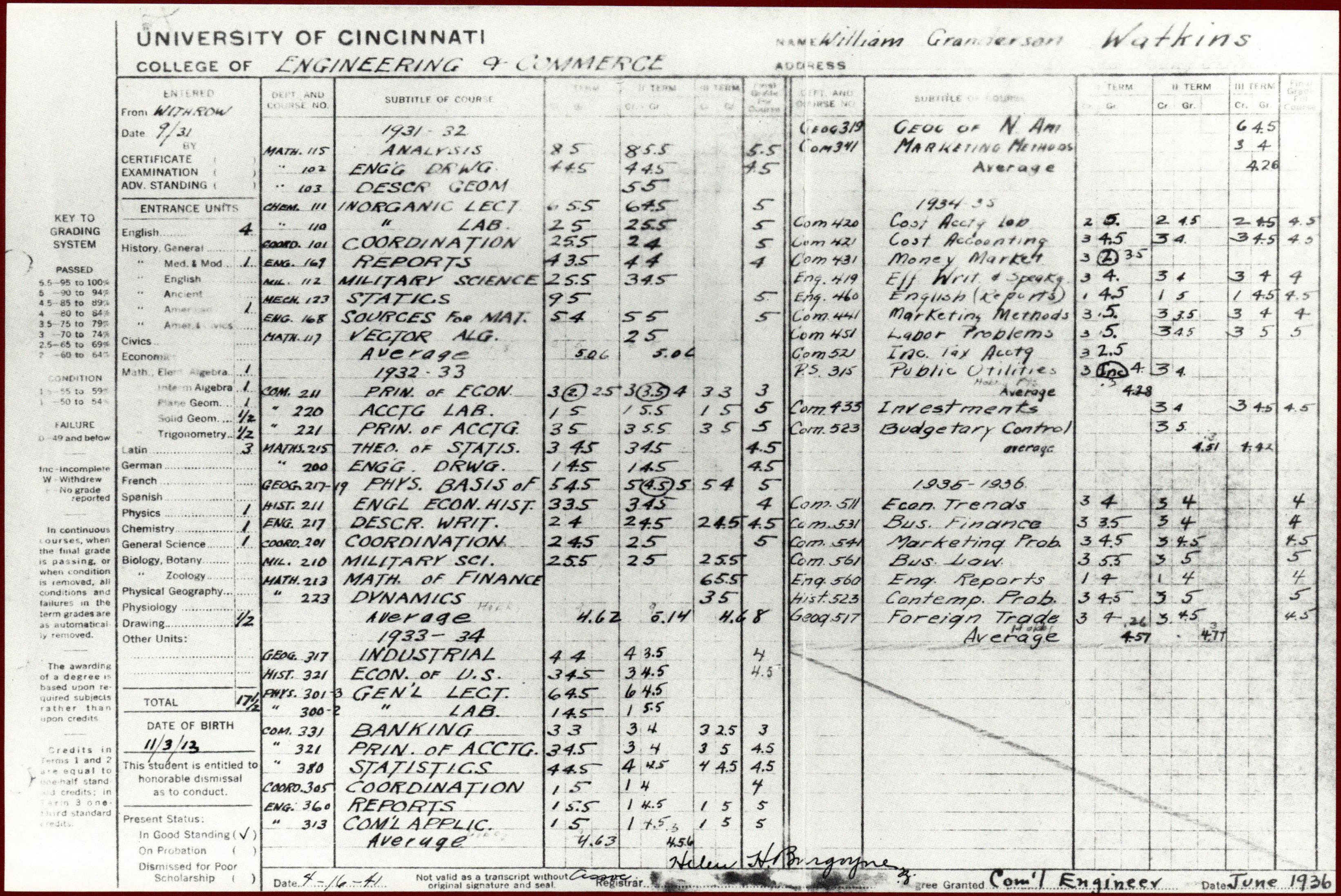
UNIVERSITY OF CINCINNATI

COLLEGE OF ENGINEERING & COMMERCE

NAME William Granderson Watkins

ADDRESS

KEY TO GRADING SYSTEM

PASSED
5.5—95 to 100%
5 —90 to 94%
4.5—85 to 89%
4 —80 to 84%
3.5—75 to 79%
3 —70 to 74%
2.5—65 to 69%
2 —60 to 64%

CONDITION
1.5—55 to 59%
1 —50 to 54%

FAILURE
0—49 and below

Inc—Incomplete
W—Withdrew
—No grade reported

In continuous courses, when the final grade is passing, or when condition is removed, all conditions and failures in the term grades are as automatically removed.

The awarding of a degree is based upon required subjects rather than upon credits.

Credits in Terms 1 and 2 are equal to one-half standard credits; in Term 3 one-third standard credits.

ENTERED From WITHROW Date 9/31 BY CERTIFICATE EXAMINATION ADV. STANDING

ENTRANCE UNITS: English 4; History, General; Med. & Mod. 1; English; Ancient; American 1; Amer. & Civics; Civics; Economics; Math., Elem. Algebra 1; Interm. Algebra 1; Plane Geom. 1; Solid Geom. 1/2; Trigonometry 1/2; Latin 3; German; French; Spanish; Physics 1; Chemistry 1; General Science 1; Biology, Botany; Zoology; Physical Geography; Physiology; Drawing 1/2; Other Units:

TOTAL 17 1/2

DATE OF BIRTH 11/3/13

This student is entitled to honorable dismissal as to conduct.

Present Status: In Good Standing (✓) On Probation ( ) Dismissed for Poor Scholarship ( )

Dept. and Course No.	Subtitle of Course	I Term	II Term	III Term	Final Grade
	1931-32				
MATH. 115	ANALYSIS	8 5	8 5.5		5.5
" 102	ENG'G DRWG.	4 4.5	4 4.5		4.5
" 103	DESCR GEOM		5 5		
CHEM. 111	INORGANIC LECT	6 5.5	6 4.5		5
" 110	" LAB.	2 5	2 5.5		5
COORD. 101	COORDINATION	2 5.5	2 4		5
ENG. 169	REPORTS	4 3.5	4 4		4
MIL. 112	MILITARY SCIENCE	2 5.5	3 4.5		
MECH. 123	STATICS	9 5			5
ENG. 168	SOURCES FOR MAT.	5 4	5 5		5
MATH. 117	VECTOR ALG.		2 5		
	Average	5.06	5.06		
	1932-33				
COM. 211	PRIN. OF ECON.	3 (2) 2.5	3 (3.5) 4	3 3	3
" 220	ACCTG LAB.	1 5	1 5.5	1 5	5
" 221	PRIN. OF ACCTG.	3 5	3 5.5	3 5	5
MATHS. 215	THEO. OF STATIS.	3 4.5	3 4.5		4.5
" 200	ENGG. DRWG.	1 4.5	1 4.5		4.5
GEOG. 217-19	PHYS. BASIS OF	5 4.5	5 (4.5) 5	5 4	5
HIST. 211	ENGL ECON. HIST.	3 3.5	3 4.5		4
ENG. 217	DESCR. WRIT.	2 4	2 4.5	2 4.5	4.5
COORD. 201	COORDINATION	2 4.5	2 5		5
MIL. 210	MILITARY SCI.	2 5.5	2 5	2 5.5	
MATH. 213	MATH. OF FINANCE			6 5.5	
" 223	DYNAMICS			3 5	
	Average	4.62	5.14	4.68	
	1933-34				
GEOG. 317	INDUSTRIAL	4 4	4 3.5		4
HIST. 321	ECON. OF U.S.	3 4.5	3 4.5		4.5
PHYS. 301-3	GEN'L LECT.	6 4.5	6 4.5		
" 300-2	" LAB.	1 4.5	1 5.5		
COM. 331	BANKING	3 3	3 4	3 2.5	3
" 321	PRIN. OF ACCTG.	3 4.5	3 4	3 5	4.5
" 380	STATISTICS	4 4.5	4 4.5	4 4.5	4.5
COORD. 305	COORDINATION	1 5	1 4		4
ENG. 360	REPORTS	1 5.5	1 4.5	1 5	5
" 313	COM'L APPLIC.	1 5	1 4.5	1 5	5
	Average	4.63	4.56		

Dept. and Course No.	Subtitle of Course	I Term	II Term	III Term	Final Grade
Geog 319	GEOG OF N. AM			6 4.5	
Com 341	MARKETING METHODS			3 4	
	Average			4.26	
	1934-35				
Com 420	Cost Acctg Lab	2 5	2 4.5	2 4.5	4.5
Com 421	Cost Accounting	3 4.5	3 4	3 4.5	4.5
Com 431	Money Market	3 (2) 3.5			
Eng. 419	Eff. Writ. & Speakg.	3 4	3 4	3 4	4
Eng. 460	English (Reports)	1 4.5	1 5	1 4.5	4.5
Com 441	Marketing Methods	3 5	3 3.5	3 4	4
Com 451	Labor Problems	3 5	3 4.5	3 5	5
Com 521	Inc. Tax Acctg	3 2.5			
P.S. 315	Public Utilities	3 (Inc) 4	3 4		
	Average	4.28			
Com 435	Investments		3 4	3 4.5	4.5
Com 523	Budgetary Control		3 5		
	average		4.51	4.42	
	1935-1936				
Com 511	Econ. Trends	3 4	3 4		4
Com 531	Bus. Finance	3 3.5	3 4		4
Com 541	Marketing Prob.	3 4.5	3 4.5		4.5
Com 561	Bus. Law	3 5.5	3 5		5
Eng 560	Eng. Reports	1 4	1 4		4
Hist 523	Contemp. Prob.	3 4.5	3 5		5
Geog 517	Foreign Trade	3 4	3 4.5		4.5
	Average	4.57	4.77		

Date 4-16-41. Not valid as a transcript without original signature and seal. Registrar. Helen H. Burgoyne. Degree Granted Com'l Engineer Date June 1936

This handwritten transcript represents the academic career of a commercial engineering student from 1931 to 1936. The courses, which included military science during the first two years, were graded on a scale of 1 to 5.5 and converted to a percentage (see the key at the far left). William G. Watkins was off to a perfect start, earning a 5.5 in Analysis (Math 115) in his first term (see the top entry).

LIGHT IN THE ATTIC

George Sperti, E.E. '23, directed the college's Basic Science Research Laboratory from 1925 to 1935. In his studies of ultraviolet light, he invented the sun lamp and created frozen orange juice concentrate, Preparation H®, and Aspercreme®.

All scientific disciplines are bound together by the same laws of nature, so perfectly consistent that they must have a supreme creator as their single source. That belief was the theme of the Basic Science Research Laboratory, where, for the first time at an American university, scientists in biology, chemistry, physics, and medicine explored together fundamental research questions of pure science. To start, they pursued questions of personal interest to themselves. And as their discoveries found practical applications, they planned to turn the patents over to the university, sell the rights to industry, and invest the royalties in the laboratory.

The man Schneider chose to direct the laboratory was George Sperti. He had caught the dean's attention as the undergraduate, who, on a co-op assignment with the Union Gas & Electric Company, invented a light meter to measure kilovolt-ampere demand. Sperti's personal convictions matched those of the laboratory exactly. He was a religious man who believed that the orderly laws of science were evidence of a divine intelligence. And although he was courted heavily by industry, he chose to remain at UC and devote his career to general research.

With a staff of freelance scientists, the Basic Science Research Laboratory never had a secure income, so conditions remained primitive at best. Most early equipment was handmade by the staff. The laboratory set up shop in the attic of Cunningham Hall, where in the summer heat waves rose from the floor; in the winter the staff had to dress warmly enough for the outdoors. The researchers survived the cold, but their mice developed middle-ear disease and pneumonia.

Despite the challenges, Sperti and his staff discovered how to duplicate the sun's rays with an electric lamp. One of their first thoughts to application was preserving fruits and juices with Vitamin D from the rays. From that discovery came frozen orange juice concentrate and the sun lamp. The laboratory got its first financial break when General Foods purchased rights to Sperti's patent for selective irradiation of ultraviolet light. But long negotiations with General Foods over royalties prevented the lab from resting on its laurels. The Great Depression and a fire in 1931 that gutted the laboratory compounded the financial strain.

In 1935 Sperti left UC and joined Cincinnati Archbishop John McNicholas to establish the Institutum Divi Thomae, a graduate school where scientific research was also conducted within the belief of a divine creator. At the Institutum he applied ultraviolet light to animal cells in cancer research. He discovered that extracts of those cells, which he named biodynes, meaning "life force," were effective in a burn salve. That process fostered the methods to formulate more pain relievers, which became the household words Preparation H® and Aspercreme®. Perhaps it was more than coincidence that research in light brought astounding success to the career of a scientist who was bound to the idea that all matter is part of a great unity.

When Schneider reorganized the Basic Science Research Laboratory after Sperti's departure, he introduced graduate-level cooperative education, which sent advanced students to work for companies with research divisions. Cooperative-fellowships and bequests funded the lab into the 1950s.

"The ice man had not yet been put out of business by Frigidaire and air conditioning was not yet a reality. I remember working on a tracing of an arch bridge that was one of my senior structures problems with a towel under my hand to keep perspiration from ruining the work in the heat of a Cincinnati June."
—Oscar S. Bray, C.E. '32

tory to explore the scientific principles of lithography. In response to the opportunity to be a part of the institute, printers nationwide who were interested in initiating organized research for their industry formed the Lithographic Technical Foundation. Their laboratory was the first of its kind for that industry. Robert F. Reed, a chemical engineering graduate of UC, was the first director.

Closest to Herman Schneider's heart was the Basic Science Research Laboratory, also established in 1924. It operated as part of the larger Institute of Scientific Research, but it differed from the tanning and lithographic laboratories in its freedom to investigate any research problem that looked promising to the staff.

Researchers at the laboratory came from a variety of scientific divisions, to generate creative and useful research through the interplay among their fields. They were directed by Dr. George Sperti, a UC graduate in electrical engineering. This interdisciplinary approach was consistent with both Schneider's and Sperti's belief that ultimately all scientific truth is one, and that a problem in chemistry or mathematics is a problem for all science.

Dean Schneider was not able to devote most of his attention to the institute for long. When President Hicks announced his retirement in 1928, the UC Board of Directors again asked him to assume the presidency. This time Schneider accepted, but only to be acting president while the board continued their search for a permanent successor. Schneider appointed Dr. Robert Clyde Gowdy, professor of physics, to serve as acting dean of the College of Engineering and Commerce. The board nevertheless persisted in recruiting Dean Schneider, and he finally consented reluctantly, on the condition that he could retain the title of Dean of the

Dean Robert Gowdy embellished his stern teaching style with a dry wit. When he once heard students whispering to each other during a surprise quiz, he turned off the ceiling fans and said, "That should make it easier for you gentlemen to hear."

College of Engineering and Commerce. In 1929 the board elected him president of the university, and Gowdy continued to serve as acting dean of the college. Howard B. Luther, who had been professor of civil engineering since 1922, assumed the William Thoms Chair in 1929, and also became assistant dean. Schneider continued to direct the Institute of Scientific Research and the Basic Science Research Laboratory.

"Professor Robert C. Gowdy's format seldom changed. He would stride on stage from a door behind his desk as soon as the bell rang–and lecture continually for the entire class period. He was always well-dressed and his sartorial elegance became a topic of student discussion. One day we decided to take up a collection to buy him the brightest colored tie we could find–just to see what his reaction would be. In due time, our gift-wrapped purple and lavender creation found its way to his desk. As the bell rang, our professor came through the door–started for the blackboard, saw the package, stopped, unwrapped, and held up the tie for everyone to see. He thanked us, removed his own tie and dropped it in the wastebasket, put on our tie, and without another word, cooly went on to his lecture! Needless to say, we all admired Professor Gowdy's unique reaction and it became a classic story within our group."
—Bruno C. Agostini, C.E. '40

"My first co-op job was for the Bureau of Public Roads as a member of the survey crew laying out the first, and still only, road through the Great Smokey Mountains from Gatlinburg, Tennessee, to Cherokee, North Carolina. Pay was three dollars per day."
—Robert Curtis, C.E. '39

"Robert Hundley was the best professor I ever had. In a loud voice he would say, 'Mr. Jones (slide rule in hand but figuring on a corner of the blackboard), if you're going to calculate that way, please put your slide rule back on your desk,' or 'Mr. Smith (at blackboard puzzling with a vector diagram), if I tied that rope around your waist, threw you out that open window, then threw the rest of the rope out with you, what would be the tension in the rope?' "
—Robert G. Bennett, M.E. '39

During Schneider's presidency, the university's physical and financial resources continued to expand. New facilities included a library (rededicated as Blegen Hall in 1983), Wilson Auditorium, Teachers College, the YMCA building, and on the east campus the Clinic Building and the Charles Franklin Kettering Laboratory for Applied Physiology. Braunstein Hall, the building devoted to physics instruction, was under construction at the time. The college established the Department of Aeronautical Engineering with a gift from the Jacob G. Schmidlapp Trust Fund of $25,000 over five years. Under Professor Bradley Jones in 1929, it was the first of its kind to use the cooperative system.

UC President Raymond Walters was known as Statistician Laureate of American colleges and universities. For over thirty-five years he was the national authority on enrollment statistics.

The tradition of awarding honorary degrees began in the college in 1930, initially to recognize UC graduates who had earned noteworthy recognition in their fields. The first man honored was Joseph Strauss, the 1892 graduate in civil engineering who designed and built San Francisco's Golden Gate Bridge. He was still negotiating obstacles in constructing the bridge when the university awarded him an honorary doctor of science, and described the yet unbuilt bridge as "a culminating work of beauty and masterly engineering, a triumph of engineering skill." The bridge was completed in 1937, and fulfilled the university's glowing predictions. Originally the bridge was to have a combination cantilever-suspension design. But as research in the 1920s led to more knowledge of steel, his design changed to a suspension bridge, with its four thousand-foot span being the longest suspension in the world at that time.

Back in the president's office, Schneider did not find his work satisfying. "I don't like this job," he complained to a reporter for the *Christian Science Monitor.* "I'd rather be the merest cub in the research laboratory than to be here." In his first year as president, he notified the Board of Directors that he did not believe his health was strong enough to remain very long in the office. It was clear he wanted to be dean again. So the board appointed a committee, which included Schneider, to seek a new president.

The college bestowed an honorary doctorate of science on Joseph Strauss, an 1892 civil engineering graduate, for designing and constructing the Golden Gate Bridge. The original university building on Vine Street was being razed just when construction of the bridge commenced. Strauss sent for a brick from the building, and before workmen began pouring concrete for the bridge, he ceremoniously placed the brick in the anchorage so that "the traditions it represented would be preserved for all time."

Still, it would be two more years before a new president took office and Schneider could go home to the dean's office.

It was Schneider who was in fact responsible for the initial contact that led to recruiting a new president. In January of 1932, Dr. Raymond Walters came to Cincinnati to speak to the Association of American Universities. Schneider heard the speech, and later telegraphed an invitation for Walters to return and meet with the university board. After two interviews, the board elected President Walters in March 1932.

Raymond Walters came to the University of Cincinnati after a twenty-five year career in journalism, publishing, teaching, research, and administration. He was a graduate of Lehigh University, Herman Schneider's alma mater, and was assistant to the president at Swarthmore College when he accepted the UC presidency. He was also associate editor of the journal *School and Society*, and published an annual statistical report of American college and university enrollments, which had earned him the unofficial title of statistician laureate of higher education.

From 1927 to 1953, Co-op Day was an annual event for each department to exhibit their activities to the public. The idea was suggested by Ellis Crawford (C.E. '28) in an editorial for The Co-operative Engineer. For the 1934 Co-op Day, these students displayed an experimental streamlined motor car. From left to right, they are: William Thompson (M.E. '34), Charlotte Gross (B.S.A. '34), Ned Drucker (M.E. '34), Allen Lloyd (M.E. '34) and Lea Klein (Com.E. '36). Lloyd and Thompson designed and built the car for their senior M.E. thesis.

President Walters took office near the lowest point of the Great Depression. Three years earlier, on October 29, 1929, the stock market crashed and business activity and employment across the country fell rapidly. For the 1931-32 academic year, Herman Schneider's last as president, the university froze salary increases. The first order of business after Walters' arrival was a ten percent reduction in salaries for that year. Most vacancies remained unfilled, and fellowship money went into the general university fund. Naturally, the university's income from tuition, fees, taxes, and endowments dropped in accordance with the decreased student population. Some faculty

"When I graduated, a thesis was required, perhaps to justify the E.E. degree. My faculty advisor was Professor Culver, who had been approached by the head of chemical engineering to furnish something like the equipment which was the subject of the thesis. Each year a number of theses were selected for presentation and they were judged with a prize for the winner. After the obligatory presentation of purpose and theories of the equipment I made, we had a demonstration of the actual welding. My roommate happened to have experience as an arc welder. While I described the operations, he demonstrated. (I don't believe we made it known that he was an experienced arc welder.) Of course, he had much trouble establishing and maintaining an arc without benefit of the superposed high-frequency arc, but with it, voila!"

—Robert B. Killen, E.E. '36

"I worked for Container Corporation of America for forty-six years. From the twenty cents per hour start (as a co-op student), I retired as vice president and director of manufacturing. For me, the co-op system made it possible to attend college during the depths of the Great Depression and produced a job opportunity that lasted my entire working career."

—Frank G. Jones, Ch.E. '37

It Is More Blessed to Transmit Than to Receive

During the 1920s the plaque on the wall of the electrical engineering department's amateur radio station, 8CAU, read, "It is more blessed to transmit than to receive." But during Cincinnati's Great Flood of 1937, both transmitting and receiving were blessed events, for the station was the only means of communication when city phone lines failed. In the hills of Burnet Woods, the campus was spared damage from flood waters. And at one of the highest points, on the top floor of Swift Hall, Professor William Osterbrock supervised the student operators in dispatching information to the Red Cross and WCPO, the commercial radio station. Dispatching occurred around the clock, and because they had to order all other station operators off the air, the Federal Radio Commission gave the operators a special frequency, free of any interference. Since then the station has been called W8YX.

Since Swift Hall was built, the station has occupied the top floor. With a founding date of 1921, it is one of the oldest student organizations at UC. Osterbrock, the station's first advisor, was instrumental in planning for Swift to accommodate a radio station. The towers are not merely fastened to the roof, but are built into the superstructure of the building.

As W8YX radio was communicating vital information in Cincinnati, another UC engineer was pioneering ways to make those transmissions. Modern facsimile began in the mid-thirties with a machine that first transmitted over radio waves, not telephone lines. Captain William G. H. Finch was granted the first patent for a facsimile machine, which reproduced pictures as sharp as later telephone models. The cost of an office model in 1939—about $2,500—prevented the machines from appearing everywhere, but uses for the Finch facsimile were widespread nonetheless.

After Finch transmitted a color picture of a parrot over two wires from New York to Chicago for publisher William Randolph Hearst, all major Hearst news offices were equipped for facsimile communication. For Hearst he also installed the first international radio typewriter circuit between New York and Havana. He developed a compact machine to fit in an automobile glove compartment for police officers to transmit photographs and fingerprints.

In Cincinnati, Powell Crosley bought early licensing rights for Finch's machines at WLW radio. Finch himself was a pioneer in commercial radio, having put several stations on the air, including WGHF in New York City. He invented two home machines; the first received two-column newspapers transmitted from an equipped radio station, but it was made obsolete by television. The other, marketed for children,

All communication came from UC during the Great Flood of 1937. For two weeks straight, around the clock, Professor William Osterbrock, above, and his students at radio 8CAU dispatched information to all parts of the city. The amateur radio station was the city's only means of communication during the worst phases of the flood, when the student operators ate and slept in shifts at Osterbrock's house. In the fifties his son Carl studied electrical engineering at UC and also became a professor in the department.

"The memories of the 1937 flood stay vivid in the minds of my era. It was eerie to see a town the size of Cincinnati come to a halt in every way. Yet it was beautiful to see people help each other and face the problems that arose."
—John Franklin Schneider, M.E. '41

In 1939 Captain William Finch was at the New York World's Fair with his daughter, Eloise, to see the exhibit of his facsimile machines. The machines were displayed by International Telephone & Telegraph, one of many corporations which bought licenses to operate Finch facsimiles in the thirties.

read comic strips from Hearst magazines. When the comics were fed through a spring-driven tape unit, a voice from the machine sounded the words on the strip.

Finch attended the College of Engineering between 1915 and 1920, and though he never earned a B.S., he did earn a law degree in 1923 from Columbia University, not to be a practicing attorney, but to protect his own patents—all 208 of them. His first invention, in 1913 at age fifteen, was a ship's radio equipped with an ink recorder to pick up wireless signals. His inspiration was the Alexander wireless bill, passed in the early 1900s, which required all ships with fifty or more passengers to employ a wireless operator.

He conducted some early experiments in fax transmission as a co-op student at Bullock Electric Company in Norwood, which later became the Allis-Chalmers Corporation. By July of his freshman year, Finch was licensed by the government to operate a five-hundred watt radio transmitter downtown at 15 East Third Street. Not long after, he was operating a small boat in the Ohio River by remote control from a water tower fifteen miles away in Cheviot. As a co-op student at the Cleveland Electric Illuminating Company in 1917, he conducted carrier current communication experiments over the company's high voltage lines between Cleveland and Akron. After organizing Finch Telecommunications to manage his commercial inventions, Finch continued his electronics career with the Navy. During World War II, while he was in charge of electronic warfare, the Federal Communications Corporation used his circular antenna to detect German spy operations in the U.S.

The Finch Gallery of Electronic Communications, in Baldwin Hall, was dedicated by Finch himself in 1990. It is a permanent display of early electronic equipment from the 1930s and World War II. Finch endowed the display and donated the equipment, which he obtained during his years in the Radio Club of America.

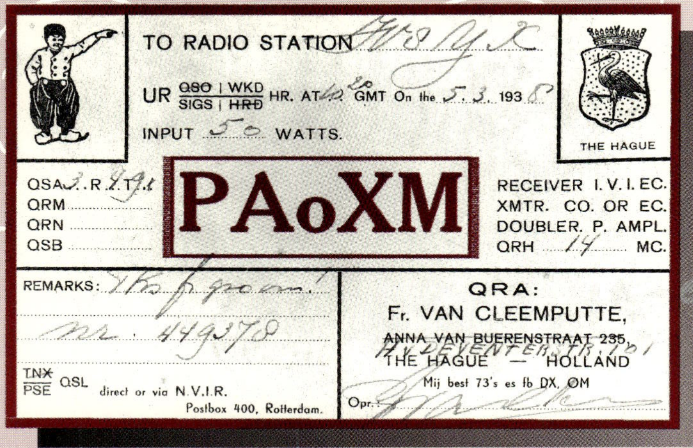

Amateur radio stations exchange QSL cards as a way to acknowledge a successful communication. Over the years, W8YX has sent and received thousands of cards to and from all parts of the world. This card, from a station in Holland, was sent to UC on March 5, 1938. The abbreviated writing in the remarks column means, "Thanks for your card, old man!" Old Man is the official title radio operators use to address each other.

"During the school year 1933-34, heavy rains caused flooding in the Cincinnati area. The UC power plant in the lower section adjacent to the stadium had flood waters high enough to immerse the power generating equipment. Professor A.M. Wilson, head of electrical engineering, assigned Carl Gieringer (a second year graduate student and assistant instructor) and myself (in the first year of the same program) to see what we could do to help restore equipment to operating condition. Inspection indicated that no apparent damage was caused other than the potential of insulation degradation due to the water immersion. As substantiation of the practical experience background of the co-op program, Carl proposed that the generator-motor windings be connected in a transformer arrangement with an adjusted load to cause current flow heating of the power equipment, thereby effecting the necessary drying out. It worked, saving the University many dollars of replacement cost."

—John E. Keto, E.E. '32, M.S. '35

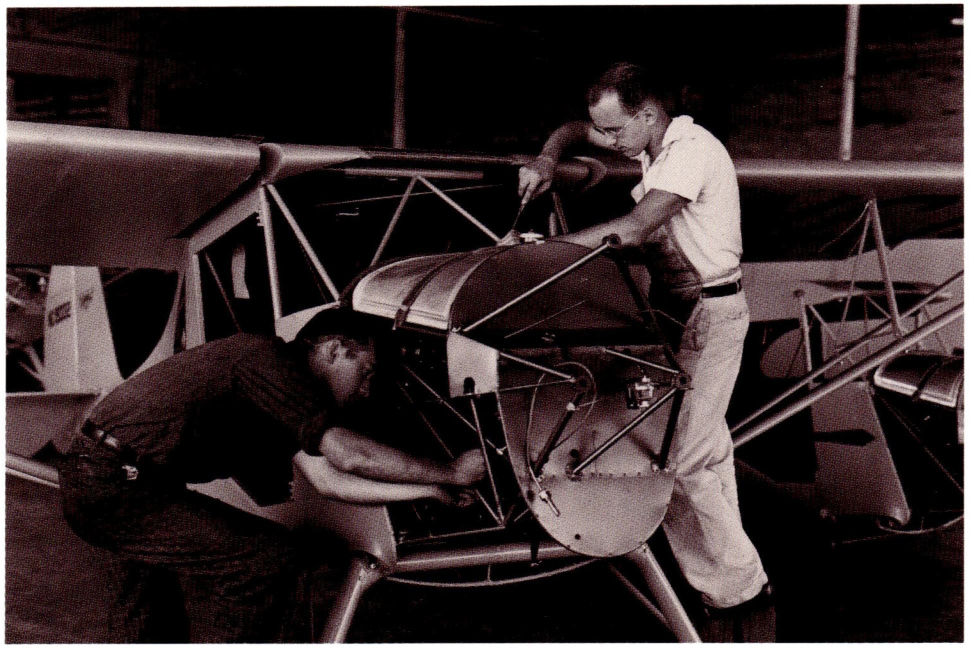

The co-ops performed airplane assembly work at Aeronca (Aeronautical Corporation of America) at Lunken Airport in 1936. At the left is Raymond Bisplinghoff, A.E. '40, who became professor of aerospace engineering and deputy head of that department at the Massachussetts Institute of Technology. At the right is A.C. Ballauer, A.E. '38, who became a project engineer in the Helicopter Division at McDonnell Aircraft Corporation in St. Louis.

took leaves of absence offered to them at half-pay for a semester, but virtually all were able to retain their positions.

The Great Depression provided the definitive test for cooperative education, although the program had already weathered general business recessions. Though shaken, it remained intact through the depression. One reason it survived is that students were expected to learn about business from the bottom up and often worked as machinists, laborers, electricians, and clerks. This practice turned out to be a real boon to the program while the job market was devastated.

But the depression did not leave the program unscathed. As enrollment declined, it was still impossible to place every student in an appropriate job. In 1932 only half of the eligible students held co-op jobs, and the numbers steadily recovered from that low point. By 1933, seventy-five percent were employed, and in 1934, ninety percent were again at work.

Faculty coordinators went to extraordinary lengths to find students jobs, sometimes any job, even babysitting. Students who could not get work were permitted to continue their education by attending classes year-round. Students with no other sources of funds received help from the National Youth Administration in the form of employment at the university.

The college altered the cooperative schedule in 1934 to accommodate regulations on working hours imposed by the National Recovery Act. Freshmen attended classes full-time from September until Christmas, and then began a seven-week alternation in January. To synchronize with this schedule, classes above the freshman level also went on a seven-week schedule. Because it accommodated winter and spring vacations, and

"University schooling was a serious business in 1935, a depression year. Some of my classmates were as much as ten years older than I. Some were married, and a substantial minority had worked and saved for several years in order to get started in school."

—William J. Cunningham, C.E. '40

resulted in three equal periods of work and classes, the college kept the schedule after the depression ended. When conditions began to improve in 1935, the university immediately restored a small part of the faculty salary cuts as enrollment slowly climbed.

No special celebration acknowledged the thirtieth anniversary of cooperative education in 1936, except for a small book by Schneider entitled *Thirty Years of Educational Pioneering*. That same year the college introduced a plan for cooperative education in graduate study in the Institute of Scientific Research.

In the beginning of 1939, the college had every reason to expect a continued bright outlook. The country had survived economic disaster and enrollment in the college climbed steadily back to pre-depression levels. And a new addition nearly doubled the size of the Chemistry Building, thanks to the Works Progress Administration.

But the college would soon be shaken by a personal loss. On Tuesday, March 28, 1939, shortly after 8:00 a.m., Herman Schneider died of a heart condition at age sixty-six. He had suffered a mild heart attack about ten days earlier. Robert Gowdy, who again became acting dean of the college, conveyed the shock in a May 1 letter to students: "The death of Dean Schneider . . . can hardly be more of a shock for any of you than it was to us here, for we had no warning that his condition was serious . . . Tuesday morning, March 28, I was lecturing—same old 9:30 lecture—when there was a knock at the door and a message that I was wanted on the phone . . . Dean Schneider had died. At 10:30 I went over to Baldwin Hall. The corridor was filled with students . . . The most sincere and touching tribute I have ever seen paid to any man was expressed in the attitude of the students as they stood silently in small groups or alone,

Half of the engineering corps involved in building Mariemont, Ohio, were UC co-ops. For the city's planners, employing the students was more feasible than a full-time staff, because the flexible co-op force could be increased and decreased easily as construction tasks fluctuated.

"My first co-op job was with the Cincinnati Butcher Supply Company as a laborer. The president of the company was Mr. Schmidt, who used to walk through the plant every morning. I decided that I wanted Mr. Schmidt to know that Joe Mancini was working there, so every morning as he would pass me on his stroll, I would speak up and say, 'Good morning Mr. Schmidt' in a very timid voice at first. Mr. Schmidt would not even recognize that I had spoken, so as time went on, I used to speak louder and louder until it got to the point where I was really shouting. The best answer from him that I could get was a grunt. But from that day on, when he came through the plant I would holler, 'Good morning Mr. Schmidt!' and would receive a grunt in reply.
I at least knew that he knew I was there. Later on, after I had graduated, I was offered a scholarship from the Cincinnati Butcher Supply Company for Harvard Business School. I could not accept, but I often wondered if the offer of the scholarship was because of my speaking out so loudly each day as Mr. Schmidt came through."
—Joseph A. Mancini, A.E. '35

"It was just a few months before my graduation in June 1939 when we lost Dean Schneider. I had had the good fortune to talk with him on several occasions during my student days, and I had the honor of delivering the undergraduate eulogy at the campus memorial service. It was as heartfelt as I could make it."

—**Donald J. Montgomery, Ch.E. '39**

stunned, bewildered, by what had happened . . . I have not taken the Dean's place . . . For the time being I happen to be 'carrying on.'" Gowdy did become dean, and would continue to "carry on" through the coming war years.

Herman Schneider was a national figure, and his death was marked by extended obituaries in the *New York Times* and the *Chicago Tribune*. Locally, he was honored as a hero. At the funeral, Cincinnati Mayor James Garfield Stewart headed the list of honorary pallbearers. Active pallbearers included President Walters, students and faculty from the college, and representatives from professional societies. The university dismissed classes for the day.

At the time of his death, more than twenty-five schools and colleges of engineering, including ones in Russia and China, were successfully using Schneider's cooperative concept of education. Joining theory and practice together into a working system of education, while simultaneously building ties between "town and gown," were only the first accomplishments that drew on this concept of unity.

Schneider was also successful in building ties between the commercial and technical facets of business, and in connecting art with utility for manufacturers. His clear and comprehensive vision led the college to a prominent position in engineering education.

"Dean Herman Schneider was very genial and friendly to all students. He advised us not to cram for exams, but to go out and have an oyster stew the night before, an activity he considered relaxing."

—Anna Bartruff, Ch.E. '32

War, Peace and Progress

CHAPTER 4
1939-1966

The Army Specialized Training Program prepared soldiers on college campuses across the country for specialized duties. The 1555th Unit commenced at UC in 1943 and trained over two thousand soldiers. After one year, at the height of World War II, the army discontinued the program at UC and sent more soldiers directly into combat. On March 10, 1944, the 1555th Unit marched its farewell parade in front of McMicken Hall.

WHEN HERMAN SCHNEIDER died unexpectedly, Robert Gowdy was promoted to the dean's office. Since it was not Schneider's management style to delegate responsibility, there was little provision for others to take over in his absence. Although Dean Gowdy claimed to be "carrying on" instead of taking over, he also became a distinguished college leader. He steered the "engine school" through financial drought when the student population plummeted and led the college's response to the war effort in the tradition of innovation established by Schneider.

Gowdy's first years as dean, before the war, were calm and productive. In his first report to President Walters, he noted that "there have been very few changes in the faculty and no significant modification of faculty or department organization. The continued heavy teaching load under congested conditions has been met by a fine spirit of co-operation on the part of the staff."

Student life gave way to military regimen during the war years as most students enlisted with ROTC, ASTP, or CADET, the women's auxiliary. Raising the flag in front of McMicken Hall are Robert Tour, Ch.E. '42, Dick Tullis, Bus. '47, and Wilburn Pean, M. E. '42. Tour was a student in chemical engineering, Class of 1942, and the son of Reuben Tour, professor of chemical engineering.

"Reuben Tour and his thermo course were legendary in chemical engineering and Earl Farnau was an unbelievable fount of knowledge. I recall going into Professor Farnau's office right after lunch to ask him a question and leaving there, weary and hungry, at 7:30 p.m. He had been talking all that time about not only the question I had asked but about every possible ramification of it."
— Frank S. Riordan, Jr., Ch.E. '40

Gowdy also noted with satisfaction a finished addition to the Chemistry Building and graduation of the first class of commercial engineering students who entered the college as freshmen, rather than as transfer students.

As the economy improved under depression recovery measures, so did the health of the co-op program. When George Sperti left the Basic Science Research Laboratory in 1935, Dr. Walter Soller became director of its graduate programs, sponsored primarily by the Stephen Wilder Endowment Fund, in which students spent alternate six-month periods in the laboratory and at industrial research locations. The laboratory had graduated from the attic of Cunningham Hall to a wing in Hanna Hall, then later to its own building behind McMicken. It was renamed the Applied Science Research Laboratory. So successful was this program that the board of directors of the university authorized a Graduate Department of Applied Science in 1940. It acquired the functions of the laboratory and offered the degrees of Master of Science in Applied Science and Doctor of Science, with co-op experience. The Lithographic Research Laboratory and the Tanners' Council Laboratory then became independent units as well.

The long list of memorials to Dean Schneider began when the engineering quadrangle was dedicated the Herman Schneider Quadrangle on Co-op Day 1940. John Wessel unveiled his commissioned portrait of the late dean on this day, too.

Even though the United States had not yet entered the war, by 1940 college faculty were already in service to national defense. For the United States Office of Education, Dean Gowdy studied the educational needs of the Cincinnati industrial area for strengthening defense activities in the region. For the National Defense Research Committee to the Council of National Defense, he surveyed the technical research facilities and personnel of the university. Mathematics professor Louis Brand surveyed the university's resources in physical science and mathematics for the council. Bradley Jones, department head of aeronautical engineering, served on the Civil Aeronautics Bureau with a group of experts who prepared an enlarged civilian pilot training program. After studying modern gauge technique at the Watervliet Arsenal, Professor Reuel Smith helped develop courses

Calculating equipment occupied much desk space in the statistics lab of the 1940s. Yet when this photo appeared in the 1942 yearbook, the caption rightfully hailed these comptometers as a far cry from the abacus.

"An engineering student was easily recognizable on campus because he carried a ten-inch Keuffel & Esser slide rule, either a log-log trig model or a log-log decitrig model. There was another way to distinguish most engineering students in those days—we usually were seen in slacks and either leather or suede jackets rather than in suits, which were more common in the other colleges of the university."
—Thomas W. Chappelle, M.E. '41

"Like many of my classmates after the war, I was married and anxious to become a breadwinner as soon as possible. Consequently, we all elected to forego the co-op experience in order to complete our academic studies as soon as possible. Little did we realize that in 1947 the opportunities for employment as new aeronautical engineering graduates would be very bleak indeed. Nevertheless, some of us were fortunate enough to obtain a position upon graduation."
—Leonard Segel, A.E. '47

of instruction in the design and use of gauges in the mechanical engineering curriculum. In anticipation of war, the Reserve Officers' Training Corps (ROTC) enrollment in 1940 at UC had doubled from the previous year.

For the University of Cincinnati, as for the rest of the country, all aspects of life changed when the United States entered World War II.

The war effort created an unprecedented demand for professional engineers, which continued long after the forties. The American Bridge Company offered jobs to all graduates of the civil engineering class of 1941, as did the Ordnance Department to all chemical engineers meeting physical requirements.

While there was plenty of work for everyone, on campus there was almost no play. Fraternities assumed inactive status as their houses became military barracks. Athletics limped along with a scarcity of players; only the fencing and basketball teams played regular seasons, and all athletic facilities accommodated military drills. The students postponed their extracurricular activities, intercollegiate sports, and proms for the duration of the war. The Cincinnati Auxiliary Defense Effort and Training recruited over three hundred women students, and ROTC recruited a total of nine hundred students.

Academic conditions changed as well. The university accelerated the academic programs and granted degrees to students in absentia. Fewer and fewer students enrolled at UC as qualified men of fighting

The electrical engineering faculty that taught on campus during the war was a small group that got smaller. On the top row from left to right are Alexander B. Bereskin and Albert C. Herweh. On the bottom row are Lawrence Culver, Alexander M. Wilson, and William C. Osterbrock. Wilson was in poor health, Culver left the university in 1942, and with virtually no students to teach in 1944, Bereskin took a year's leave to join Western Electric. Herweh and Osterbrock were the remaining staff.

Patriotic photos promoting the war effort came from everywhere, including the Goodyear Girls program. These two engineering co-eps, who worked on airplane engines for General Electric, added smudges of oil to their cheeks for effect, but forgot to dirty their hands a bit.

"Students who left [for war] as immature adults, and who experienced combat, returned as mature adults who would not be inhibited by a professor. If a person had faced a German machine gun fire under combat, the combative instructor held no fear. The instructor was the one who had to change or run the risk of being intimidated. Needless to say, it was a learning experience for both student and instructor."
—Richard J. Reiman, E.E. '49

age enlisted in the armed forces. By September 1943, enrollment dropped from 1,765 to 427 in the College of Engineering and Commerce, and in the School of Applied Arts from 256 to 181. As faculty also left for active military service or specialized civilian assignments, the college accelerated its programs by eliminating vacations and co-op terms. The seniors of 1942 and 1943 graduated two months early. In 1944 they graduated eight months early, in October instead of June, and became the Class of 1943½. In February 1943, 79 high school juniors were admitted to the college a year early as freshmen, on the recommendations of their high school principals.

This class of ASTP soldier-students was taught history by Edward Hake Phillips (A&S '40). At the end of their session, when the army discontinued the 1555th unit, Phillips was moved to take a picture of the men, many of whom, he feared, would not return home.

The college contributed to the war effort primarily by educating both soldiers and civilians. The students and faculty who remained on campus were joined by soldiers who came for special training. Women students trained through certificate programs to qualify for war-related work, and by the autumn of 1943, they were for the first time admitted to all departments of the college.

The School of Applied Arts offered a one-year program in mechanical drawing to its seniors. The Department of Aeronautical Engineering and the Goodyear Aircraft Corporation trained "Goodyear Girls" as airplane factory supervisors. A twelve-week program trained inspectors for the Ordnance Department. A five-month training course for the Signal Corps Junior Engineering Aides provided personnel for Wright Field in Dayton, Ohio.

Work in the research laboratories was almost exclusively war-related. In the Department of Chemical Engineering Professors Reuben Tour, Hoke Green, and Harold Garber were contracted by Wright Field to carry out Project MX121. For two years they conducted preliminary searches for suitable chemicals in rocket fuels. Their calculations were tested on a simulated rocket, fired indoors in their Chemistry Building laboratory lined with sandbags.

The chemical engineering department also conducted research in the ultra-

The Herman Schneider Foundation was established by the Engineering Society of Cincinnati in 1943. The Schneider Foundation sponsored programs in engineering, math, and science in local high schools. The foundation also maintained this house on McMillan Street and Woodburn Avenue, which served as the Engineering Society headquarters for twenty-seven years.

THE ARTIST WHO DISCOVERED BENADRYL®

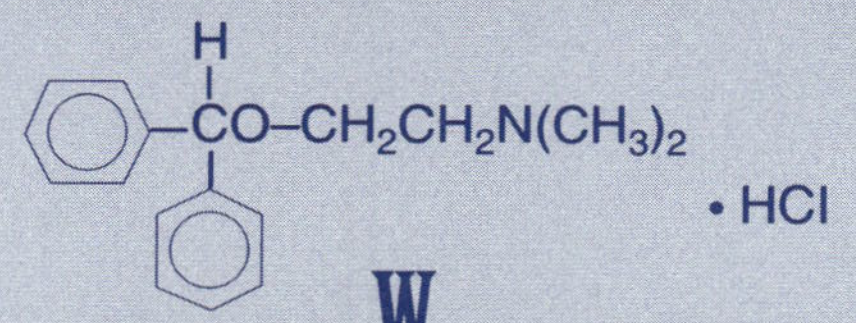

Were it not for the Great Depression, George Rieveschl, Jr., might have made a career of drawing advertisements for Benadryl instead of discovering it. The antihistamine drug was developed by Rieveschl, in a joint effort with Dr. W. Frederick Huber, when Rieveschl was a faculty member in the College of Engineering during the 1940s. But his original ambition was art, not chemistry.

He was born in Lockland, Ohio, in 1916. Instead of enrolling in a traditional high school in 1929, he went to Cincinnati's Ohio Mechanics Institute, which later became the OMI College of Applied Science at UC. He graduated from the commercial art program in 1933, at the height of the Depression, when there were no jobs for artists. But Rieveschl had long been an amateur chemist with his own basement laboratory, where he even made his own paints for his art courses. So he then enrolled at UC, commuting from Lockland, and in three and a half years completed a bachelor's degree in chemistry. Encouraged by Professor Frank Ray, Rieveschl earned his master's in 1939 and a doctorate in chemical engineering by 1940.

But again employment prospects were bleak. After two hundred letters brought no job offers, he accepted a position in the chemical engineering department. He quickly rose in the academic ranks working with graduate students on research projects to determine the chemical structures of anesthetic and antispasmodic compounds.

On a 1943 trip to Detroit to visit a former student at Parke-Davis Pharmaceuticals, he arranged for the company to test the Rieveschl-Huber research. After three years of testing, Benadryl was released commercially in 1946. It was licensed as an antihistamine, and also used as an antispasmodic in the preparation of cough medicines. Further applications have relieved skin rashes and have been effective in treating emotionally disturbed children.

George Rieveschl, Jr., (B.S. Chem. '37, M.S. Chem. '39, Ph.D. Ch.E. '40, D.Sc. '56) discovered the compound for the drug Benadryl as a faculty member in the College of Engineering.

During the years of testing Benadryl, Rieveschl joined Parke-Davis as a research chemist, where he again rose in the ranks to become vice- president of commercial development. Joining Parke-Davis, however, did not mean leaving UC for good. In 1970 he returned and began a third career as the university's vice-president of research and development. In 1973 he endowed the annual Rieveschl Award, which recognizes faculty achievements in scientific and scholarly research. Rieveschl Hall, in the Brodie Science and Engineering Complex, was named for him in 1987, in recognition of his years of service to the university and to the field of scientific research.

"My good friend Manny Augenstein and I would walk downtown to the Wheel Cafe on Saturday nights to get a good meal and save the ten cents on streetcar fare. We would walk back through the Over-the-Rhine district, and almost always see a fight, or somebody thrown out of one of the bars. We never were even remotely concerned about our own safety on any of these walks; and we saved ten cents each time!"
—Charles K. Hodell, Jr., A.E. '48

drying of oxygen when George Metzger, owner of the General Air Conditioning Corporation in Oakley, approached his friend Dean Gowdy in his search for an independent research site to develop a drying system for the air force. The air force approved the university as an impartial researcher, so Professor Garber and Professor William Licht developed a dessicant which Metzger's company sold to the air force during the war.

Scientists at the Tanners' Council Laboratory developed a shoe leather for the military and civilians, which was twenty-five percent more durable than all earlier types of leather. The Applied Science Research Laboratory was contracted by the Rubber Reserve Company for research in synthetic rubber.

When he became dean in 1946, C. Albert Joerger was the first to head the College of Engineering after it ceased being the College of Engineering and Commerce and returned to its original name.

But the college's focus was on teaching and training. The campus became a military institution when the Army chose UC as a site for its Specialized Training Program (ASTP), which used university resources to prepare soldiers for specialized military functions. The 1555th Army Specialized Training Unit commenced in February 1943, enlisting eight hundred students in programs across the university. The College of Engineering taught pre-engineering and basic civil, mechanical, and electrical engineering courses for the ASTP. The army leased Memorial Dormitory and eight fraternity houses as barracks. McMicken, Hanna, and Cunningham Halls became dormitories, and professors moved their offices and personal libraries almost anywhere. Peak enrollment reached 2,450 soldier-students, and over half were in engineering programs. But shortly after the first class graduated in the spring of 1944, the army cut the program back by half, to send more soldiers directly into combat.

Diversions from the war centered around more tributes to Dean Schneider. In 1943 the Engineering Society of Cincinnati established the Herman Schneider Foundation "as a center for the development and advancement of the ideals and accomplishments of the distinguished educator and scientist, Herman Schneider." The society was founded as the Engineers' Club of Cincinnati in 1888, a professional organization uniting all technical societies in Cincinnati. The

This 1940s photo of the secretaries in the College of Engineering represents years of continuous university service since 1911. From left to right are their names and the years employed at UC: Florence Stapleton (1927-1972), civil engineering and secretary to Dean Francis Bird, College of Business Administration; Margaret "Barbara" Kiessling (1926 to 1955), registrar's office; Louise "Winnie" Winston Brown (1931 to 1942), registrar's office; Anna Teasdale (1911 to 1952), secretary to Deans Schneider and Gowdy; Freda von der Halben (1925 to 1964), assistant to Miss Teasdale; Janis Lyons Focke (1932 to 1968), secretary to Deans Joerger, Justice, and Wandmacher; and Marie Seuberling Ludeke (hired in 1940 and still employed at the time of publication), math department secretary and later assistant to UC President Henry Winkler.

"My first year in Cincinnati, I lived in the economy dormitory located in Nippert Stadium. It was operated for freshmen men at a charge of fifty cents each week. Those living in this dorm were selected on the basis of high school scholastic record."
—Robert C. Nolte, M.E. '43

Schneider Foundation bought a house on McMillan Street and Woodburn Avenue, to serve as headquarters.

Foundation funds bought and maintained the building and sponsored local high school programs in engineering, science, and math. Inside the building, with the meeting and reception rooms, dining room, and staffed kitchen, was a memorial library devoted to technical subjects. The first president was Frederick V. Geier, president of Cincinnati Milling Machine, and UC's President Walters was a trustee.

Another memorial to Schneider was his biography, *Ambassador to Industry*, published in 1943 by Clyde Park, an English professor in the college. These activities gave the college a temporary focus away from the war, but in the meantime more and more students left school for the military. After the War Department discontinued the ASTP, virtually all male students became subject to active duty. The 1944-45 academic year was perhaps the most critical, when enrollment dropped to forty-five men and one hundred eighty women in the college, and the loss of income created a financial crisis.

To balance the deficit for the entire university, the administration launched the Emergency Fund Campaign, for which Frederick Geier served as lead corporate sponsor. In a show of support that amazed even those who had initiated the fund drive, he pledged $100,000 from his company, one fourth of the campaign goal. Other companies followed his lead, and raised a total of $406,000, so the university could retain all faculty and staff.

When the war ended, several events prevented a complete return to business as usual. Dean Gowdy resigned in 1946 because of illness and Professor C. Albert Joerger, head of the Department of Mechanical Engineering, became the new dean.

"As a commercial engineering student, my courses and goals were business-oriented. I had been working for Montgomery Ward as a home furnishings buyer for five years when I decided to start Whiting Manufacturing Company in 1949. Montgomery Ward was sorry to see me go, but they ordered two thousand comforters from me before I even had a plant."
—Carson Ross Whiting, Com.E. '33

Army barracks housed married students attending UC through the GI Bill. While the husbands and fathers attended college, their families—of as many as five—lived side by side in close quarters. Even closer were the individual household quarters, which were thirty-six feet long by eight feet wide.

Outside forces shaped much of Joerger's tenure, just as they had Gowdy's. With the war's end, Joerger faced the enormous task of rebuilding the faculty and educational programs while managing the flood of veterans returning to college under the Servicemen's Readjustment Act, better known as the GI Bill. The immediate priority of processing the large number of these students consumed Joerger's first years as dean.

"Professor Howard B. Luther was head of the civil engineering department and taught a course in advanced statics called Special Problems. He had little patience with students who blindly followed theory without using common sense. His favorite question was, 'How would you approach this problem if you didn't have the disadvantage of a college education?' "
—Joseph G. Wiener, C.E. '42

"Professor Bradley Jones loved to illustrate some of the basic principles of aerodynamics through humorous examples. One which stands out in my mind is his illustration of the importance of the fact that in the theoretical equation which predicts the distance required for an aircraft to take off, rho (ambient air pressure) appears in the denominator of the equation. Professor Jones would tell a story about a publicity stunt sponsored by a cereal manufacturer back in the early thirties. The cereal manufacturer had the brand name of his cereal painted in large letters on the side of an old Boeing passenger plane. He loaded the plane up with photographers and publicists and landed it on a high plateau in the Rocky mountains. After all of the picture taking aimed at showing all of the great things you could accomplish by eating this cereal, it came time to take off. A broad knowing smile would spread over Brad's face, as he asked the rhetorical question, 'What do you think happened when they tried to take off? Well, I'll tell you what happened. That airplane is still sitting on the top of that mountain—there wasn't enough room to take off because those guys forgot that rho is in the denominator!'"

—Ernest Ratterman, A.E. '51

As the three divisions in the College of Engineering and Commerce and the School of Applied Arts grew and expanded, they became more independent of each other. In 1946 each became individual colleges: the College of Engineering under Dean Joerger, the College of Business Administration under Dean Francis Bird, and the College of Applied Arts under Dean Ernest Pickering. Within this new structure, the Department of Coordination in the College of Engineering became the university-level Department of Coordination and Placement, directed by Henry Calvert Messinger; and the Department of Chemical Engineering became the Department of Chemical and Metallurgical Engineering.

A deluge of students flooded the campus. Enrollment nearly doubled in the first autumn after the war, and each following year, the population broke records. By 1947 the 19,000 students campus-wide were almost thirty times as many as the 664 who were enrolled at the height of the war in 1944.

In addition to housing married students on University Avenue, army barracks provided classroom space behind Braunstein and Swift Halls and in front of Baldwin Hall to accommodate the overflowing enrollment in the late forties.

For the first time in history large numbers of older students attended college full-time. Prior to the war, adults attended evening classes, but the war created a new class of students who went to college with support from the GI Bill. To accommodate the married students with families, the army constructed two-story barracks buildings along University Avenue and next to Memorial Dormitory. The complex, called Vetsville, was a community of students, their wives, and children. The quarters were tight; each unit, thirty-six feet long by eight feet wide, was partitioned to accommodate two families. Barracks buildings also made for temporary classroom space north and south of the quadrangle.

The workload created by so many students taxed the faculty and staff, yet the professors found it rewarding to teach veterans. The war had matured them, and they were interested in their studies and serious about completing their degrees. At the request of the campus Veterans' Office, the college waived co-op requirements for the veterans and married students who already had work experience. For the rest of the students, the cooperative schedule returned to a seven-week alternation of two sections in a fourteen-week term, with three terms making up a school year. Classes still met six days a week, with students attending class forty hours each week.

Enrollment in the college was at a record high, yet requests from companies for co-ops began to outnumber the qualified students to fill positions. Before World War II, the program stressed learning from the bottom up on work assignments. Students worked in menial and entry-level technical positions, especially when jobs were scarce during the Depression. Also, as unions emerged in industry, manual labor on co-op jobs virtually disappeared. So in the late forties more students began their co-op experiences with actual engineering work.

Post-war engineering enrollment peaked in the 1947-48 school year with 2,154 students, but by the 1950-51 school year the engineering population decreased to 1,488 students. The campus at this time looked as much like a military institution as it did during the war.

In addition to housing students in Vetsville, barracks went up as extra classroom space behind the quadrangle on the north and south sides, and right in front of Baldwin Hall.

Another change in appearance came in 1950 when McMicken Hall, the original home of the College of Engineering, was razed and rebuilt. It was a beloved landmark, but the interior needed such extensive repair that the board decided to replace the building. The redesign, which incorporated 650,000 bricks from the first building to support interior walls, profoundly changed the Clifton Avenue view of the campus; its colonial facade with a Christopher Wren spire incorporated the separate Hanna, McMicken, and Cunningham Halls into one building.

Redesigns and permanent changes took place inside the buildings too, as the importance of advanced training for engineers ascended with the role of technology in political and economic power. Engineering students had been earning graduate degrees from the Department of Chemical and Metallurgical Engineering and the Graduate Department of Applied Science for years, but the necessary goal for the college in the 1950s was to offer graduate instruction in every department.

A joint project among the College of Engineering, the Engineers' Council for Professional Development (ECPD), and local industries was an early step toward that goal. In 1952 the ECPD, a council of eight national engineering societies, selected Cincinnati as the pilot city for the First Five Years program, which supported the continuing personal and professional development of young engineers in their early years of employment. One objective of the program was to provide an opportunity for young professional engineers to pursue graduate degrees without leaving their jobs.

"Robert (Hot Rod) Hundley was one of the best teachers we had. We got a quiz problem every day. Each person had a different problem which was passed out on 3x5 cards. If our quiz grades averaged 90% or above we didn't have to take the tests. Unfortunately, if a person had a 90% or above on the last day before the test, his card for that day had a nearly impossible problem (by pure chance, according to Professor Hundley). I was the first one in the row and we would pass the cards down to other students. On the last day before a test, I learned to take the bottom card (although he instructed us to take the top card). He knew I did this (after the fact) but he couldn't say anything without admitting that he stacked the deck. Each person in my row saw me take the bottom card, so they did, too. The last guy in the row was really upset because he ended up with my card several times and almost flunked the course."
—Max Brown, M.E. '59

"Late one night I was burning late oil studying when a Sigma Chi fraternity brother came in from a late date and asked me if I smelled smoke downstairs. We decided to call the fire department since we couldn't find the source. With thirty fraternity brothers sleeping on the top floor at 1:30 a.m., all was quiet. When I called the fire department and requested that they come to investigate, I said not to make any noise because thirty people were asleep on the third floor. Needless to say I was the joke of the fraternity and campus for several days. The fire department came roaring down University with everything they had—ladders, trucks, etc.—and reported to the newspaper that I had asked them to be quiet because thirty people were trapped on the third floor."
—John Z. Dozier, M.E. '42

PLASTICS PATENTS BY A HUSBAND AND WIFE TEAM

One significant development in engineering to emerge during the 1950s was materials science. It expanded over the course of the following twenty years to produce plastics which replaced metals in many phases of many industries. Richard Francis (Ch.E. '34) played a significant role in this technological revolution. His thirty-year devotion to plastics research furthered industries as diverse as naval defense and sporting goods.

Francis began his career in reinforced plastics in the Newark, Ohio, laboratories of Owens-Corning Fiberglas. In 1940 he moved to Toledo to take a research position with the Plaskon Division of Allied Chemical. At that time Plaskon had developed a new resin which could be laminated with applied pressure a hundred times less than existing resins. After using Plaskon's new resin, the navy added a plastics division to its ordnance laboratory and recruited Francis as director to initiate production of its materials.

In 1944, Francis returned to Owens-Corning to help develop fiberglass products and to design a shop for making prototypes for field evaluation. The prototypes included a T-6 aircraft wing and an auto with its body and structural parts made of reinforced plastics. While working as a consultant to American and European companies, he contributed significantly to developing glass fishing rods, GlasShafted golf clubs, the fiberglass antenna, and glass arrows. In 1946, Francis received two individual Naval Ordnance Development Awards, one in recognition of his earlier work in developing fibrous glass reinforcements for plastics, and the other for developing methods of molding products of reinforced plastics.

The career of Richard Francis may well have been confined to his research were it not for an event in his life, quoted by a 1964 Columbus Dispatch article on memorable events in Ohio: "That day Dick walked into the Walgreens drug store, where I was working in 1948, was really love at first sight." It was this chance meeting that brought together the creative plastics technologist and the acumen business mind of Clara Hare, who married Francis in 1953. Together they built Francis Industries in Pataskala, Ohio.

They began in 1956, working unassisted in three hundred square feet of space. By 1968 their plant occupied 8,500 square feet for ten employees and their production tripled from the previous year. Clara, affectionately called The Boss by everyone, especially Richard, used to cook in the company kitchen for her employees in the early days, before management consumed all of her time.

Francis patents for fiberglass products include high-strength rods, tubes, pipe, fishing rods, canoes, golf clubs, arrows, and CB radio antennas. Their research included naval weapons, corrosion problems, and wave propagation in the communications field. Since their research put them beyond the range of conventional equipment, they built and designed their own—even their factory roof, so that it could accommodate their research in radio wave propagation. They developed fiberglass communication antennas by combining early research in sporting goods with work in fiberglass insulation for Argonne National Laboratory's atom smasher, a 12.5 billion electron volt proton accelerator.

Their partnership took on its greatest challenge in 1970 when Richard, despite a terminal illness, made every effort to continue his research, and Clara, acting as executive, employer, wife, and nurse, managed the business. After Richard's death in 1971 Clara continued as chief executive officer and increased production of Francis Industries by one thousand percent prior to selling the business in 1977.

Richard and Clara Francis designed and built all of their own research and production equipment. Richard (Ch.E. '34) is running the drill press in this 1968 photo of the Francises in their machine shop.

Adapted from "A Biography of Richard J. Francis," prepared for Clara Francis by Joseph A. Brannon

Fourteen local corporations, led by Ernest S. Fields, executive vice-president of Cincinnati Gas and Electric Company, joined the university in sponsoring the pilot project. Civil engineering professor Cornelius Wandmacher chaired the college's involvement, which guaranteed a three-year cycle of course offerings, and the industrial sponsors agreed to subsidize the courses as necessary. Enrollment of three hundred fifty students in the first year surpassed all expectations, and the industrial subsidies were never needed.

By the mid-fifties the college was once again prosperous through a large, yet manageable enrollment of 1,250. After the immediate concerns of depression, war, and veterans' education were over, the university could once again acknowledge significant events with some extravagance. First came President Walter C. Langsam's inauguration in 1955. The former president of Gettysburg College succeeded Raymond Walters in a show of pageantry which the previous two presidents had forfeited. Three hundred delegates from around the world, in full academic regalia, witnessed the ceremony which was recorded on film.

The banner event for 1956 was another celebration, the fiftieth anniversary of cooperative education, also organized by Professor Wandmacher. "Education and Industry at Work for Progress" was the theme of a week-long conference and exhibition, from April 19 to 25. But it was more than just a celebration. The event was jointly sponsored by UC and local industries to provide a forum for new ideas on how to perceive and utilize cooperative education. The celebration did this in three ways: it reviewed the industrial progress of the past half-century, it emphasized advanced education and training as the necessary resources for meeting the challenges of the future, and it acknowledged achievement in the science work of high school students to attract them to technical and industrial fields.

President Walter C. Langsam's inauguration in 1955 was a large and festive occasion, one which the previous two presidents had forfeited.

The events of the first day of the conference, Herman Schneider Day, commemorated the first fifty years of cooperative education. The Herman Schneider Quadrangle was officially dedicated on this day, although the board of directors had named it in 1940. During the ceremony, the quadrangle was enclosed

"One nice spring day in differential equations class, my interest level slipped until I was asleep in the first row. The professor told the fellow next to me, 'Wake up Hochadel.' That woke me up enough to hear the student respond, 'You wake him up–you put him asleep!'"

—Jack B. Hochadel, M.E. '56

DEMONSTRATING PROGRESS AND ENGINEERING TO VISITORS WITH ALCHEMY, MOUSE TRAPS, AND MAGIC MACHINES

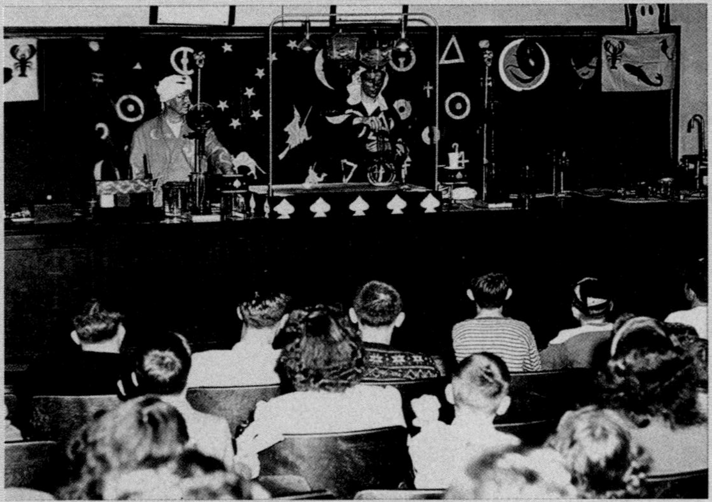

For thirty years, Edward Schlesselman donned a wizard costume to perform "A Night in Alchemy." His magic show was a perpetual favorite among Co-op Day visitors.

Ellis Crawford, a chemical engineering student, Class of 1928, entered the Cooperative Engineer's essay contest in 1927, won first place, and inspired a twenty-six-year tradition called Co-op Day. His essay argued for an Engineer's Day as a "means of drawing the attention of men in and around Cincinnati." For the college, he envisioned "a better and closer feeling among the students . . . [and] a spirit of friendly rivalry between departments to present the best and most novel exhibits." The administration agreed with the idea, and in 1928 the cooperative colleges opened their doors to the rest of the university and to everyone else in Cincinnati.

On Co-op Day, the students and faculty in each department hosted exhibits and lectures, enlightening their visitors to the different fields of engineering, and the merits of the co-op system. Seniors displayed their thesis projects, and many students created projects especially for Co-op Day. In 1928 a group of electrical engineering students imitated popular cartoonist Rube Goldberg, who drew machines of fantasy chain reactions. The students' machine, called "Waking up a Co-op Student," was completely functional and attracted so many visitors—including Goldberg himself—that the parts were nearly worn out by the end of the day.

Exhibits often reflected the technology of the times. At the 1941 Co-op Day, visitors were informed of the ways the college could contribute to technical training in national defense. In 1950 many people were attracted to a novel machine which recorded their voices and played back the sounds. It was called the magnetic tape recorder. The most enduring attraction was "A Night in Alchemy," a magic show produced by Edward Schlesselman, the supervisor of laboratory equipment for the chemistry department. Schlesselman dressed as a sorcerer and demonstrated the principles of chemistry to high school students, in the hope of attracting them to science studies in college. He changed water to wine, concocted clouds of smoke, and ignited candles spontaneously. "A Night in Alchemy" became the most popular attraction, as Schlesselman performed it every year, for the twenty-six-year duration of Co-op Day.

The ultimate open house event was in 1956, the year of the college's fiftieth anniversary of cooperative education.

On Co-op Day 1947, Dale Whitford (A.E. '51), fourth from the right, wearing the light jacket, watches a demonstration of the water flow table built by the senior aeronauticals. For the demonstration, an airfoil was placed on the glass table and the thin film of water flowing over the surface simulated supersonic flow by using polarized light. One could actually see the shock waves.

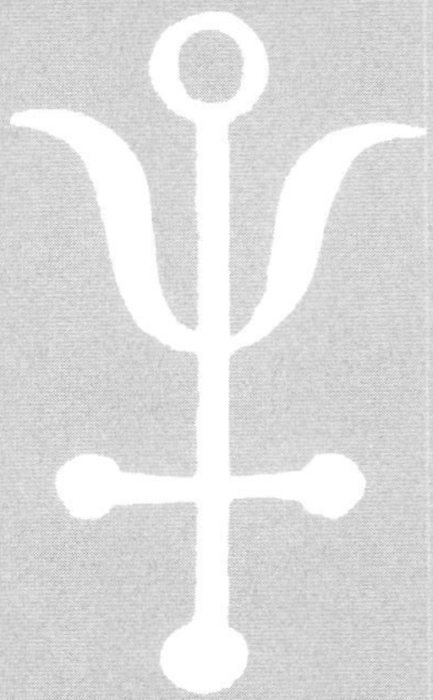

Antimony

Thousands of people visited the "Panorama of Progress" exhibit in the Armory Fieldhouse every day during the week-long celebration of the fiftieth anniversary of cooperative education.

Co-op Day was just one part of the week-long celebration in April of that year. The largest event was the "Panorama of Progress," an exhibit of seventy cooperative companies from across the country. In the Armory Fieldhouse they displayed their individual scientific and industrial achievements since 1906, the founding year of cooperative education. Over 53,000 students, alumni, and local citizens of all ages attended "Panorama of Progress." Like Co-op Day, the "Panorama of Progress" was an event which families could attend together and to which schoolteachers brought their classes.

Another program drew hundreds of high school students to campus on Saturday mornings. For six weeks every year, the college faculty presented a "Saturday morning program," the idea of chemical engineering professor Saul Arenson. He began the program in the 1930s, believing it made good business sense for the college to give away samples of its products to potential customers. So he contacted high school teachers from all over—Cincinnati, Dayton, northern Kentucky, and Indiana, asking them to nominate their best science students to attend the presentations.

Chemical engineering professor Saul Arenson initiated the Saturday morning program for high school students because he believed it made good business sense for the university to give samples of its product to potential customers.

Hundreds of students flocked to Wilson Auditorium to watch science demonstrations by the college's most captivating faculty. Physics professor Isay Balinkin performed light shows to demonstrate how the eye perceives color. He also demonstrated atomic fission by setting off a chain reaction with one hundred mouse traps and as many rubber balls. Geology professor Otto von Schlichten took the students on a slide show tour of the Grand Canyon and chemical engineering professor Robert Delcamp made foam rubber in front of the students. Chemical engineering professor William Licht assumed Arenson's duties in 1940, and Professor Delcamp ran the program from the mid-forties to the late fifties, when it ended.

Co-op Day, the "Panorama of Progress," and the "Saturday morning program," served many purposes. They did promote the college, but they also gave the community a look at the work of engineers. Before television, these events revealed to people some methods of an industrial society, the properties of the natural world, and how engineering bridges the two.

Zinc

"In those days, 1935, UC offered Saturday morning lectures in Wilson Auditorium to interested high school students who wanted to learn about different possible fields of study. I remember especially the dramatic program put on at one of those Saturday morning sessions by Professor Isay Balinkin of the physics department. He made physics look and sound almost too exciting for my taste. Other lectures I remember were an impressive geological tour of the Grand Canyon and an interesting lecture on aeronautical engineering by Professor Bradley Jones."
—Edward Hake Phillips, A&S '40

Dean Howard Justice was a civil engineering co-op, Class of 1919, who lettered in football, basketball, and track. He was such a valuable player that when he had to drop out of school for a semester to earn more money, the University News published an unsuccessful plea that the Ohio-Conference allow Justice to remain with the Bearcats. In this 1951 photo he was an assistant dean and director of admissions for the college, greeting visiting seniors from Ross Township High School. From left to right are Justice, Priscilla Smith, Beverly Kanaus, Norma Shaw, and James Rader.

potential of the faculty. The Research Council published a twenty-page booklet, "Staff Publications," listing significant writings of faculty members in engineering, mathematics, and the physical sciences from 1953 to 1957. The committee recommended that the college recruit a prominent research scientist to cooperate with the Graduate School of Arts and Sciences and the University Research Foundation, in an effort to expand graduate work and research in engineering and science. They hired Hans Ernst, a long-time director of research for the Cincinnati Milling Machine Company, to be the first Herman Schneider Professor of Research. Ernst started the first cooperative doctoral program in the country in September 1959 with the Graduate Cooperative Fellowship Program (GCFP). The GCFP was a means for professional engineers to take graduate courses while working on basic research problems. The program was funded by an initial $200,000 grant from the Ford Foundation, and other industry-sponsored fellowships.

The GCFP differed from undergraduate co-op in that the graduate fellows pursued individual programs tailored to their particular careers in industry. Through the GCFP these fellows were able to retain their jobs, instead of severing connections with employers to earn graduate degrees. The program became yet another mutually beneficial relationship for the university and local industry, since industrial businesses were a part of the race for technical progress in America, and engineers in industry as well as academics needed advanced training. Most of these students attended the Evening College, which by then had an enrollment of 1,100. The GCFP provided over fifty fellowships and lasted until 1965, as each department became more able to offer its own graduate programs.

Expanded graduate programs meant that faculty members who did not have an advanced degree needed to enhance their own academic credentials to qualify them to teach graduate students. Over the next several years, UC faculty members took academic leave to earn doctorates, as part of a national drive funded by the National Science Foundation. At UC, board chairman Renton Brodie set up the Renton K. Brodie

Science and Engineering Doctorate Fund in 1967. The fund supported faculty members who were completing work for doctorates in science and engineering.

Every department expanded during these years. In the Department of Civil Engineering, John David Eye developed undergraduate and graduate programs in sanitary engineering. In 1959 five students received M.S. degrees in civil engineering with a specialty in sanitary engineering. A year earlier the first undergraduate co-ops began their assignments in that program. In 1960 William Licht, head of the Department of Chemical and Metallurgical Engineering, introduced graduate programs in nuclear engineering and materials science. A few years earlier, the department began to offer the first course ever in polymer chemistry, taught by Robert Delcamp. And the Department of Aeronautical Engineering became the Department of Aerospace Engineering in 1961.

"Professor Reuel Smith wrote theses in red ink on our exams, so we gave him five gallons of red ink when we finished the course."
—George W. Kramer, C.E. '54

The resources for the expanded laboratories in these new programs came mostly from a campaign to acquire the equipment as gifts or at nominal charges. In addition to manufacturers and government agencies, the engineering alumni responded generously to the dean's request. In the 1960 issue of his annual December letter to alumni, Justice asked the help of all alumni in the position to locate valuable and useful material, and included in the letter a list of needed equipment. In his 1961 letter he announced that several graduates had procured new and surplus equipment for the college.

Dean Justice was also making widespread curriculum changes in accordance with the recommendations of the American Society of Engineering Education (ASEE). In 1955 the ASEE's Grinter Report recommended that all undergraduate engineering curricula include basic courses in each of the six engineering sciences and a sequence of humanistic-social courses. The changes made by Justice reflected the constructive action taken, in response to the report, by schools across the country. In September 1959 the revised requirements were phased into the UC curriculum. Basic instruction in mechanics of solids, fluid mechanics, thermodynamics, heat transfer, electrical science, and nature and properties of materials was now required for every engineering undergraduate.

The plans for a book of the history of the College of Engineering began with Dean Cornelius Wandmacher, who maintained meticulous files of thousands of documents relevant to the college's history. In this 1958 photo he was associate dean, in charge of advising co-op students and maintaining relations with industry.

Grants from General Electric and the Charles F. Kettering Foundation funded a voluntary non-credit course called Humanities for Engineers, which supplemented the new humanities and social science course requirements. In the program, which began in February 1961, students who were on the

"After official graduation ceremonies at the UC Armory Fieldhouse, the 1957 class of civil engineering graduates held a party at the home of one of the graduating classmates, Deane Ivers. The party lasted all evening, and into the dawn hours of the next day. Head of the civil engineering department at that time was Professor Cornelius Wandmacher, later to become dean. Wandmacher taught the senior class structures course and promised to give a technical book as a prize to the student developing the best structures design of a third-floor bridge connecting two of the quadrangle buildings. This was years before the skywalks now prevalent in downtown Cincinnati. Somehow, the book prize was never awarded. About 3:00 a.m., with all members of our class still actively partying, it dawned on us that Professor Wandmacher had not awarded the prize. At 3:15 a.m. a phone call was then made to Wandmacher, in which he was chastised for breaking his promise. Wandmacher apologized, and politely hung up after wishing us all success in our future careers. We thought the episode was concluded. But, at 4:15 a.m., the Ivers's doorbell rang. Standing at the door, in his academic robe, mortar board, and tassel over his pajamas, holding the prize, accompanied by his wife, was Professor Wandmacher! He had gotten up, and traveled all the way across town in the middle of the night. Wandmacher solemnly awarded to the winner, Charley Meyers, the book prize for structures design."

—William F. Klosterman, C.E. '57

work phase of the co-op calendar chose literature books to read and discuss in groups lead by humanities professors from the College of Arts and Sciences. Enthusiastic response to the program, which could accommodate only a fraction of students who volunteered, was a small part of the decision for the entire university to assume a common calendar, which would happen in 1963.

But before that, Howard Justice stepped down as dean in 1962, so that he could once again teach full-time before retiring. Cornelius Wandmacher succeeded him. Wandmacher had a flair for building the college's image, as he demonstrated with the fiftieth anniversary of cooperative education. As dean he continued to use his sense of ceremony to initiate traditions. The five hundred freshmen who entered the college in September 1962 were welcomed by the administration at a convocation in the quadrangle. The following spring, on May 23, 1963, was the first annual Recognition Day, where all departments could unite to recognize their outstanding seniors. On that day the first of more than thirty departmental and college awards were given to deserving students.

At the next year's ceremony, achievement in co-op work was first awarded with the Herman Schneider Medal. It came from a bequest in the will of Schneider's widow, Louise Bosworth Schneider, who died in 1947. Her bequest provided for a college scholarship fund and for a medal, awarded annually to one senior, selected from a group of finalists representing each department, who

The Engineers' Ball was the annual event where, in the 1950s, popular dress for men was a Bermuda shorts tuxedo.

"has taken fullest advantage of the unique opportunities of a cooperative engineering education." In later years, more awards recognized outstanding teaching and research.

In 1963 electrical engineering professor Alexander Bereskin began a five-year process of building one of the first university microelectronics laboratories in the nation. Bereskin was already ahead of his time, having taught courses in transistors for several years. For the laboratory, he began with grants from the National Science Foundation and UC to design and build a three-room clean area in the basement of Swift Hall. He supplemented the grants by soliciting gifts from alumni and friends in industry. Graduate students built clean benches, fume hoods, water purifiers, and compressed air and process vacuum systems. The laboratory came together slowly; by 1968 it was completely operational and funded by additional grants from RCA and Cincinnati Milling Machine.

When Cornelius Wandmacher was dean, he welcomed the incoming freshmen each year at a convocation in the quadrangle.

Bereskin would redesign the entire laboratory himself when the department relocated to Rhodes Hall in 1970.

In September 1963 the freshman class began courses on a new all-university calendar, and in January 1964 the upperclassmen followed. For the first time, co-ops were on the same schedule as the rest of the student body. Because they had followed their own calendar, the co-ops had been isolated, in a sense, from the rest of the university. Non-engineering courses were taught to them by "captive" faculty, teaching only within the college, on the co-op calendar. Therefore, electives from the College of Arts and Sciences were simply not an option because the two calendars did not mesh. It was also difficult for co-op students to participate in sports or clubs, since practices and meetings conflicted with work schedules. In the new quarter system, students began taking electives outside the college and participating in extracurricular activities, with co-op assignments extending to an entire ten-week term. Students no longer shared jobs with alternates, and began spending their entire freshman and senior years on campus.

Another change discontinued the long-standing practice of awarding the professional degree of engineer (A.E., C.E., Ch.E., E.E., M.E.), which acknowledged the extensive practical experience students obtained through their co-op assignments. As the legal process of examination and registration to be a state-licensed professional engineer became prevalent, "engineer" became a less appropriate designation for an undergraduate degree. At the same time, the undergraduate curriculum was becoming more scientific and the percentage of students continuing in advanced programs was increasing. So the college administration and the Class of 1964 decided together for the university to award a bachelor of science in the designated engineering field as the undergraduate degree.

The first business to emerge from research in the college was the Structural Dynamics Research Laboratory (SDRL), funded initially by a 1965 air force contract. For the contract, to research vibration problems in machinery, the Department of Mechanical Engineering built a mobile laboratory facility inside a

"Professor Herman Newman would occasionally tell an off-color story to some of the co-ed classes. Some of the girls in the class who did not appreciate those stories got together and planned to get up and leave the class en mass the next time an off-color story was told. The very next day Mr. Newman started to tell a story about a trade exchange agreement that involved the US sending a number of 'ladies of the evening' to England to practice their trade. He announced a boat was docked at Newport to pick up a number of girls for the trip to England. With that, the girls in the class, on cue, got up and started to leave. 'Hold on a minute, girls,' announced Mr. Newman, 'the boat isn't leaving until tomorrow!' " —William R. Smeal, M.E. '63

"As a freshman engineering student on the fourth floor of French Hall dormitory, my roommate usually retired around 11 p.m., while all others studied until midnight or 1:00 a.m. One early morning, everyone in our wing decided to make my roommate think it was actually 7:00 a.m. and time to go to breakfast at the student union. So everyone went to the bathroom and started showering and shaving, while I set our alarm clock ahead so it would ring six hours early. When the alarm rang we got up and went to shave, etc. I purposely lagged behind, so my roommate finished ahead of me and left (before daylight). Needless to say, it was 1:00 a.m. and no automobiles were in the commuter parking lot. He went all the way to the student union and found all doors locked. When he returned to the dorm, everyone had gone to bed and turned out the lights. I reset our alarm clock to the correct time. When he came into the room, I pretended to awake and asked him what he was doing up so early."
—David W. Kuhn, B.S.Ch.E. '66

van, which was the first of its kind in the world. When the air force contract was over, the vibration analysis team continued to travel to industrial sites with the van, to consult companies nationwide.

One year after that department initiated the SDRL, it introduced an undergraduate major in industrial engineering. The major grew from a program in industrial management, established in 1946, after business administration was no longer a part of the College of Engineering. When that happened, the degree of commercial engineer was replaced by the industrial engineering program, which required a minimum of two years of engineering courses.

The college began a long-standing relationship with the General Electric Aircraft Engine Business Group in 1966, when faculty from the Department of Aerospace Engineering and Engineering Mechanics began to participate as instructors in the company's Advanced Course in Engineering. General Electric initiated that program in New York City in 1923, to provide advanced training to its employees. In Cincinnati the program evolved into the ABC Program, where GE employees attend graduate school at UC.

All of the changes in the college—in graduate education, research, and industrial ties—crowded the space in the Schneider Quadrangle buildings. Dean Wandmacher oversaw plans for a new science and engineering complex, which would more than double engineering's space and radically change the appearance of the campus. Groundbreaking for the new building would coincide with the university's sesquicentennial celebration. With a new building funded and graduate programs growing steadily, the college was positioned to continue expanding, especially in new areas of research.

Civil engineering courses in the Evening College proliferated in the 1950s, along with courses in every other department as graduate education in engineering became more important in industry and education.

CHAPTER 5
1966-1986

Linking Education, Industry and Research

The ground for Rhodes Hall was broken on October 25, 1968, by Dean Wandmacher, middle, and President Langsam, right. In honor of the university's sesquicentennial, they turned the earth with a team of mules, in the way it would have been done one hundred fifty years earlier.

BY THE MID-SIXTIES THE GRADUATE student enrollment had grown immensely because more engineers from industry began returning to school for master's and doctoral degrees. In 1967 graduate enrollment had reached over two hundred full-time and six hundred part-time students, as each department in the college was by then able to offer a competitive graduate program.

Much of the graduate student population came from industry, as professionals took advantage of the latest development in cooperative education between the college and those local industries which sponsored the advanced training of their engineers. The success of the graduate programs, though certainly a boon to the entire university, was not without problems. As the student population explosion campus-wide showed no signs of waning, the administration faced some arduous decisions, including ways to obtain the necessary funds to sustain the university's growth. An obvious

A TIMELY RETURN

HOW UC's SUN-CLOCK GOT BACK ON ITS PEDESTAL

✳ Alumni who were on campus during the mid-fifties were well familiar with the new sun-clock that had been erected in the engineering quadrangle. It was a gift of the Class of 1955 and had been designed by two chemical engineering students, Ron Rosensweig and Hans Mueller. ✳ Weighing about thirty-five pounds, the sun-clock's concentric rings had been skillfully formed from brass bars. Special difficulty encountered in its construction was finding a lens with a focal length of sixteen inches that had the desired three-eighths inch diameter. Ingeniously, the problem was solved by a quick trip to a dime store where a pair of cheap reading glasses was purchased. By cutting the oversized glasses under water with a pair of scissors, the innovators fashioned their lens, which they adeptly fitted into the analemma ring. The analemma, a graduated scale shaped like a figure eight that shows the sun's declination and the equation of time for each day of the year, was skillfully chiseled by hand. When the sun-clock was finally completed, it was oxidized, treated and lacquered for weather resistance and lasting beauty. Unlike ordinary sundials that only approximated the time, the sun-clock was unique in that it could tell time within one minute's accuracy. ✳ Unfortunately, this novel invention was doomed to a precarious life. A series of mishaps and thefts plagued the clock from the time it was first set on its pedestal. In 1962, the entire sun-clock was stolen by a few inexperienced thieves. They weren't very clever in covering up their trail, and the clock was recovered a few days later. "We worried a lot about the safety of the clock when it was reinstalled," said Jim Wenner, then Director of UC's Physical Plant. "It was a very curious device, and being made of brass it had a rather high resale value. To attempt to burglar-proof it, we made certain small changes so that nothing less than a blowtorch could get it off its base." ✳ But their confidence was soon shaken, for shortly after its restoration, the sun-clock became a target for vandals. First, its lens was smashed. The Physical Plant replaced it, only to have it smashed again. And again they had to replace it. There were no further incidents until March, 1968, when all the sun-clock's moving parts, including the two brass rings, vanished one day. Everything but the base was gone—and this time, no culprit could be found. At the time, it seemed there was no way of replacing the precious instrument. There were no funds to cover the high replacement cost, and in any case, the two students who had designed it had long since left the Cincinnati area to seek their fortunes elsewhere. None of the original blueprints had been left behind. Talk about replacing the instrument was stepped up when a few interested members of the Physical Plant got together to make a determined effort to find a way to raise the money for a new clock. Several class presidents of the "post-theft" years were asked if they would be interested in a sun-clock, but always they declined in favor of something else. "Finally we decided to make an effort to replace the clock ourselves," said Wenner, who resigned as Director in 1968, to become Utilities Coordinator of the Physical Plant, "even though we didn't know where the money was going to come from." ✳ Their first step was to consult the alumni records and find out just where the the two clock creators had gone. This turned out to be the easiest task of all, for the former UC students' most recent addresses were uncovered. "It wasn't long before we heard from Rosensweig, who had since become Dr. Rosensweig, and was president of the Ferrofluidics Corporation in Burlington, Massachusetts. He wrote that he'd be glad to help. He said he would have to redesign the instrument entirely because he no longer had the original plans." ✳ In 1955, when Rosensweig and Mueller first designed the clock, they looked around for an engineering company who could manufacture the intricate apparatus. The two at last found that the Rief Manufacturing Company in Fairfax, Ohio, would undertake the manufacture. Its cost—$500. Rosensweig finished his new drawing and sent them for an estimate to the company that had built the original clock. The price had increased considerably in the fifteen years gone by. ✳ In July of 1970, the new sundial was carefully mounted on its pedestal. Since it is the tradition that every sundial should have a motto, a brass plate affixed to the pillar reads, "There is Time for Those Who Pass." And once again, when students are late for class, they can call on the excuse, "Sorry, but the sun was behind a cloud." ✳

This story is reprinted from the October 1970 issue of UC Alumnus.

To use the sundial, turn until the lens projects an image of the sun on the analemma. The image is about one quarter inch in diameter and should be centered on the line. There are two possible settings, one on each side of the analemma; the season of the year determines which side is correct. The image of the sun marks the correct month and the approximate day. When the sun is centered on the line, the time to the nearest minute can be read from the hour circle. The index for this reading is on a plate at the lower north end of the instrument.

source was the state coffers, so in 1967 the University of Cincinnati became a state-affiliated, municipally-sponsored institution, by consent of the city's voters. State affiliation provided the university with necessary funding to assure a viable future, without sacrificing the school's autonomy or its ties to the city.

An advantage of such affiliation was participation in the state-funded expansion of higher education throughout Ohio in the sixties. That support helped the University of Cincinnati expand its physical space, which meant a new complex of four buildings for the science and engineering departments. The money for these buildings was also funded by a capital gifts campaign, launched in conjunction with the university's Sesquicentennial Celebration in 1969. It was exactly one hundred fifty years before that, at its founding in 1819, when UC held its only previous intensive capital fund drive. The drive in 1969 procured over thirty-four million dollars from almost thirteen thousand groups and individuals. The funds designated for the College of Engineering were added to government grant money to build an entire new facility. It was to be the first new space built for the college since an addition to the Chemistry Building was completed in 1939.

Celebrations of this new engineering building began on October 25, 1968, with the groundbreaking ceremony. In the spirit of the university's sesquicentennial, the ground was scraped with a plow pulled by a team of mules, the way it would have been done one hundred fifty years earlier. Mayor Eugene Ruehlmann joined President Langsam and Dean Wandmacher for the event. It was later that school year, on March 21, 1969, at the cornerstone ceremony, when President Langsam announced that the name of the new building would be Governor James A. Rhodes Hall, in honor of his support for higher education.

At the ceremony, which the governor attended, a time capsule was placed in the cornerstone of Rhodes Hall. Among the documents that went inside the time capsule were commemorative programs of the sesquicentennial, current issues of *The Cooperative Engineer* and the *News Record*, faculty photographs from 1963 and 1968, college bulletins and student guides, and President Langsam's remarks naming the building Governor James A. Rhodes Hall. At another ceremony in November, the

Governor James Rhodes attended the March 21, 1969, cornerstone ceremony for the new engineering building. He complimented the university on its many achievements from 1819 to 1969, and gratefully responded to President Langsam's announcement that the new facility would be named Governor James A. Rhodes Hall.

"Robert Price's course in physical chemistry provided the theoretical foundation for chemical engineering. I remember the exams he gave that challenged even the best students. He wrote the exams on the board and as we all struggled to work the problems, he stood at the side of the room, slide rule in hand, and solved the entire exam in a few minutes."
—Thomas L. Englert, B.S.Ch.E. '67

The three living deans attended the groundbreaking ceremony for Rhodes Hall in 1968. From left to right are Dean Wandmacher, Dean Emeritus Howard Justice, Janis Lyons Focke, secretary to all three deans, and Dean Emeritus Albert Joerger.

cornerstone was secured in its setting by Dean Wandmacher, Dean of Administration Ralph Bursiek, University Director of Development William Baetz, and John Krausser, C.E. '25, chairman of the Engineering Alumni Sesquicentennial Sponsoring Committee.

Guests at the cornerstone ceremony in March included recipients of honorary doctorates and the first recipients of the Distinguished Alumnus Award. The award was initiated on that day by Dean Wandmacher to recognize outstanding alumni who, by their achievements and stature in engineering, science, industry, business, public service, or education, have reflected an unusually large measure of credit on the University of Cincinnati and the College of Engineering.

Most students in 1969 were unfazed and unimpressed with the occasion of the university's sesquicentennial. The unrest and demonstrations for the causes of civil rights and an end to the Vietnam War were so intense, that the university administration closed the campus in May 1970 for the remainder of the school year.

The college recognized one hundred fifty distinguished alumni during the sesquicentennial year. They included George Binns, M.E. '11, the only surviving member of the first co-op class, who became a vice-president at the Cincinnati Milling Machine Company. Awards were formally presented to one hundred forty-nine alumni, with the one hundred fiftieth award symbolically retained, to represent any number of alumni who could have been recognized that year, and who would be honored in the future. Wandmacher presented those first one hundred forty-nine Distinguished Alumnus Awards at the cornerstone ceremony and additional awards at subsequent ceremonies on campus and across the country throughout that year.

The "Man And . . ." series of entertainment, educational, and cultural events lasted the entire sesquicentennial year, and involved all of the colleges in exploring education, law, life, arts, and other significant aspects of the human

"During Vietnam unrest several of us were on our way to morning classes in Baldwin Hall. We found our normal route in the rear of Baldwin was locked, along with several others. Being persistent and naturally challenged by this situation, we found a way in and made our way to our classroom, pleased with out ability to overcome this hurdle. After noticing a commotion on the quad, we learned, much to our chagrin, that Baldwin Hall was being searched for a bomb! We left in a much more direct and expedient route."
—Larry T. George, B.S.M.E. '72

"My senior year was dramatically affected by the Vietnam War and the Kent State shootings. The campus was closed, almost cancelling graduation ceremonies. The fun things were lost—Class Day, senior prank, etc.—a very disappointing end to a great experience."
—Roger A. Ketterer, B.S.M.E. '70, M.S.M.E '75

endeavor. The College of Engineering's contribution was the "Man and Technology" program, with conferences, symposia, and exhibits throughout March, April, and May of 1969. They included an international machine tool seminar, an exhibit of the space program by NASA, the annual Engineering Society of Cincinnati science exhibits of high school students, and the American Institute of Aeronautics and Astronautics conference.

On Recognition Day, May 15, 1969, the Engineering Tribunal bestowed the first Awards for Excellence in Teaching. Senior John Tomke (Ch.E.), left, congratulates Boyd C. Ringo, civil engineering; Robert Kroll, aerospace engineering; Chris Passarello, instructor in engineering science; Carl Osterbrock, electrical engineering; and B.D. Mullins, instructor in electrical engineering.

Before and during the sesquicentennial year, departmental growth and changes continued steadily. By 1967 the consulting activity at the Structural Dynamics Research Laboratory (SDRL) grew to a point where it was more of a business than an academic venture, so director Jack Lemon and several staff left UC to form the Structural Dynamics Research Corporation, the college's most successful business spin-off. Two mechanical engineering professors, Ivan Morse and Francis Tse, remained at the SDRL, which, at the time, continued to be one of the only university-affiliated laboratories in the country to train students in vibrations and modal analysis.

In 1968 the Department of Chemical and Metallurgical Engineering, the largest in the college, separated into two new departments. The new Department of Chemical and Nuclear Engineering was chaired by nuclear engineering professor James Leonard. Shortly thereafter the department began offering a bachelor's degree in nuclear engineering. The new Department of Materials Science and Metallurgical Engineering was chaired by materials science professor Michael Hoch.

In 1969 the dean created a program unit in engineering science, in light

The "We are Grateful" celebration in June 4, 1971, included a recognition luncheon in the highbay of Rhodes Hall, complete with red carpet for the occasion.

"I vividly remember working on my senior research project as the anti-Vietnam War demonstrators came through the Chemistry Building, breaking the glass windows in the doors to the laboratories."
—Theodore J. Brown, B.S.Ch.E. '69

An Astronaut Adjunct

In October 1971 the first human being to set foot on the moon came to UC to teach in the Department of Aerospace Engineering.

Neil Armstrong had returned from his July 1969 mission to the moon less than a year when he first visited President Langsam and Dean Wandmacher to discuss a possible appointment to the faculty of the College of Engineering. But instead of coming to UC at that point, Armstrong went to Washington, D.C., to head NASA's aeronautics program.

Academia eventually won out over administration when on August 25, 1971, the agenda for the UC Academic Affairs Committee included two requests from Langsam: to designate the chair occupied by the dean of the College of Engineering as the Geier Professorship of Engineering Education, and to appoint Neil Armstrong as University Professor of Aerospace Engineering, effective October 1. Armstrong continued to serve NASA in the capacity of special consultant.

At UC, Armstrong's title was that of university professor because he functioned on a university-wide, multidisciplinary basis. He participated in creating the title because it was of "paramount importance" for him to be a regular, full-time professor of maximum contribution to educational and research programs. In an interview following the announcement, Armstrong said that he had been considering an academic career for twenty years, and was looking forward to the freedom and interdisciplinary approach of his position at UC.

There was a flurry of publicity immediately after his appointment; the press even raised questions about Armstrong's lack of a doctorate. A university spokesman noted, however, that although full professors of engineering without doctorates were few, UC had never hired anyone who had been on the moon, either. The media attention soon diminished, which is how Armstrong wanted it. He always shunned publicity, to the extent that he and Wandmacher together drafted guidelines to ensure his privacy.

Armstrong settled into the relative quiet of academic life, teaching undergraduate courses. His experience as a test pilot was a great benefit to students, as he could enhance theoretical material with distinct performance accounts of various aircraft. He worked outside UC with Dr. Henry Heimlich, who crafted the Heimlich maneuver, on an artificial miniaturized human lung. He accepted appointments to the boards of directors of several local companies, appeared in public service announcements for the Boy Scouts, and chaired the 1973 Easter Seals Drive in Cincinnati.

In 1978, when the college's Institute of Space Sciences merged with UC's Institute of Engineering and Medicine to form the Institute for Applied Interdisciplinary Research (IAIR), Armstrong and Frank Tepe became associate directors, under mechanical engineering professor Ronald Huston. Activities at the IAIR would lay the groundwork for the Center for Robotics Research, the Center for Biomechanics, and the Institute for Advanced Manufacturing Sciences. Armstrong remained at UC for over eight years. When he resigned in 1979, he became an adjunct professor and in 1982 was the speaker at the university's commencement exercises.

As a college faculty member, Neil Armstrong attended the AIAA Student Chapter paper airplane contest in 1973. He demonstrated his creativity with paper and scissors, but opted not to test fly his designs.

"One of my fondest memories of student days was being a student of Professor Neil A. Armstrong. One had to plan for the unexpected when Neil Armstrong was your professor. Like on March 25, 1974, when Italian actress Gina Lollobrigida appeared after class in Baldwin Hall to take some photographs. Gina was also a successful freelance photographer and was doing a series on great people. Having already photographed Castro and Kissinger, she turned to Armstrong. Her photographs appeared in the August, 1974 Ladies' Home Journal."
—Ralph E. Spitzen, B.S.A.E. '74, M.B.A. '76

"A pocketful of colored pens was a necessity in Professor Al Herweh's DC Machines lectures if one attempted to duplicate his multicolored chalkboard art depicting motor windings. A classroom sound of the past—the intense concentration and silence during an exam suddenly being interrupted by a student's slide rule crashing onto the floor."
—Richard W. Craycraft, B.S.E.E. '70, M.S.E.E. '73

of a national movement which favored a core requirement of fundamental courses, covering all disciplines, for all undergraduates. The program grew into the Department of Engineering Analysis the very next year, and was chaired by mechanics professor Ronald Huston. Graduate courses in engineering analysis were added in the early seventies. In 1975 the department's name was changed to Engineering Science, because that term better reflected the nature of the courses, and because similar departments across the country had that name. Alvin Strauss succeeded Professor Huston that year.

These core courses were required of all engineering students until 1979, when national trends shifted again, this time toward design and application. The department disbanded and most of its professors joined the aerospace engineering department, which at that point became the Department of Aerospace Engineering and Engineering Mechanics.

Other faculty joined the mechanical or electrical engineering department.

Two of the new buildings, the Powell Crosley Research Tower and Rieveschl Hall, both to house the biology and chemistry departments, were complete in 1969. The name of Crosley Tower acknowledged the longtime generosity of the Crosley Foundation to the university. Rieveschl Hall was named (in 1987) for George Rieveschl, Jr., a UC alumnus, university vice-president of research and development, and a longtime benefactor of UC. The other two buildings—Zimmer and Rhodes Halls—were complete in 1970. Zimmer Hall was named for William Zimmer, a UC alumnus, vice-president of Cincinnati Gas & Electric Company, and chairman of the Sesquicentennial Capital Gifts Campaign.

By the spring of 1971 Rhodes Hall was fully occupied. When electrical engineering moved out of Swift Hall and chemical engineering moved out

Governor James A. Rhodes Hall was designed by the Cincinnati architectural firm Baxter, Hodell, Donnelly, and Preston. The 363,930 square-foot building cost $4,500,000. The funds came from government grants and from individual and group donors to the Sesquicentennial Fund.

of the Chemistry Building, it was the first time ever that all engineering departments were under the same roof, though the nuclear engineering program acquired old chemical engineering laboratory space in the Chemistry Building. Rhodes Hall earned national attention when *Industrial Research* magazine named the building its Laboratory of the Year for 1971. The magazine's panel of six architects and research administrators recognized the new engineering building for its architectural distinction, design, functionality, and especially its adaptability to interdisciplinary research activity.

After all the departments moved into Rhodes Hall, the college once again celebrated all day on June 4, 1971. The "We Are Grateful" program recognized the sponsors and friends of the college who made the building a reality. To commemorate the event, each of the stone plinths at the entrances of the three original engineering buildings were incised with the key words from the inscription on the Schneider Memorial Bench, which reads: "Joining Theory and Practice, Linking Education and Industry, Through Knowledge and Experience." Swift Hall bears the words THEORY and PRACTICE, Baldwin Hall bears EDUCATION and INDUSTRY, and the Old Chemistry Building bears KNOWLEDGE and EXPERIENCE. Several of the recognition plaques which hang throughout Rhodes Hall were also unveiled on that day, and the six new laboratories and conference rooms were named, in recognition of the generosity of several friends of the college.

The final recognition of the day went to President Langsam, who had announced his retirement. At his ceremony, Langsam was honored for his "extraordinary distinction as president of the University of Cincinnati, trustee of the Herman Schneider Foundation, and constant supporter of the professional activities of the engineering and scientific societies of Cincinnati." Langsam was succeeded by Warren Bennis, a vice-

"George D. Moon taught our grad class mass transfer operations. He was one of the best teachers I ever had. He knew his subject. I would simply sit and listen to him in class, understanding all. Then I would make notes of his lecture at home, from memory. He was so lucid, one did not have to take notes in class."
—Jimmy D. Kumana, M.S.Ch.E '72

Graduate students in environmental engineering identify contaminants in this 1971 photograph of the environmental chemistry laboratory.

president from the State University of New York at Buffalo and a scholar and author in management and organizational behavior.

Early in the Bennis administration, in 1972, the placement and coordination of co-op students became the function of a university-wide office, the Career Dynamics Center, which was headquartered in the Old Chemistry building. This new office consolidated and centralized the placement of the co-op students and the graduating seniors.

In Rhodes Hall, the new office and laboratory space accommodated the growing graduate programs and the professional development research of faculty, both of which were becoming more important in the field of engineering education. During the 1970-71 academic year the concept of the Basic Science Research Laboratory was revived as one source of funding for faculty research. The Herman Schneider Basic and Applied Science Laboratory —

The facilities in Rhodes Hall dedicated on June 4, 1971, during the "We Are Grateful" ceremonies recognize the generosity of several friends of the college.

The Dean C. Albert Joerger Laboratory of Mechanical Engineering recognizes the gifts of Harris-Intertype Corporation, the Schriber Company, Louis Schriber, M.E. '35, and Lloyd Whitesell, M.E. '42.

The Stenger Heat Treating Laboratory recognizes the gifts of the Queen City Steel Treating Company and the Edwin P. Stenger Family: Edwin P. Stenger, Met.E. '16, Edwin A. Stenger, Met.E. '51, and Edwin C. Stenger, Bus.Ad. '70.

The Dr. George R. Terry Seminar Room acknowledges the gift of George R. Terry, C.E. '30, who became the George A. Ball Distinguished Professor of Business at Ball State University.

The Kruse Microprobe Laboratory, in memory of Carl T. Kruse, E.E. '49, was made possible by the gifts of the Cambridge Instrument Company (Ralph H. Kruse, E.E. '16, Chairman of the Board), the U.S. Steel Corporation Foundation, and the Charles R. Frueauff Foundation, Inc. In 1989 its name was changed to the John Moteff Electron Microscopy Laboratory, in memory of John Moteff, head of the Department of Materials Science and Metallurgical Engineering from 1976 to 1987.

The Armco Metal Working Laboratory recognizes a major gift by the Armco Steel Corporation through its Research Division and a number of alumni.

The Jenkins Conference Room honors A. L. "Pop" Jenkins, Professor of Mechanical Engineering 1905-1929, in recognition of a gift by Carlton E. Brown, ME '23.

From left to right are Cornelius Wandmacher, Dean George Barbour of the College of Arts and Sciences, and Winston Kock during the celebration of the fiftieth anniversary of cooperative education in 1956. Kock came to campus for the anniversary to give an address on "Undergraduate Honors, Precursors to Success" and an organ concert in Wilson Auditorium.

Winston Kock's Creative Legacy

The Herman Schneider Basic and Applied Science Laboratory, established in 1971, is a source of seed money for interdisciplinary research. Its first director was Winston Kock (E.E. '32), a man who attributed his creative genius to his diverse interests.

As a child, Winston completed second grade and was an accomplished pianist by age four. At fifteen he enrolled at UC where his talents surfaced in varsity track, in writing music for college shows, and in chess. Blindfolded, he once took on four of his classmates in chess simultaneously, winning three of the games in ninety minutes.

Too young to be admitted to the College of Engineering, Kock first enrolled in the College of Arts and Sciences and the College of Music, and never regretted it. "A very important element in the education of young scientists," he once said, "is early exposure to a wide range of interdisciplinary studies." When he turned sixteen he transferred to engineering and his undergraduate thesis revolutionized the music world. It was the invention of the electronic organ, the first to generate sound through small vacuum tubes, rather than massive pipes. This feat put an organ within reach of nearly every family that ever dreamed of owning one. His patent for the organ was the first of over eighty that he would receive throughout his career.

After he earned a master's in physics at UC, he earned a doctorate in physics from the University of Berlin in one year, and then returned to UC to teach for a year. He later studied at the Institute for Advanced Study at Princeton and at the Indian Institute of Science in Bangalore. In 1942 he joined the Radio Research Department of Bell Telephone Laboratories to work on microwave lenses, which eventually went into Bell System's coast-to-coast microwave relay circuits. While at Bell, Kock was awarded an honorary doctor of science by UC in 1952.

He moved on to the Bendix Corporation in Ann Arbor, Michigan, in 1956, and eventually became vice-president for research. At Bendix he created the prototype of today's transistor and he directed a post-World War II sonar detection project for the navy. He left Bendix for two years, from 1964 to 1966, to serve as the first director of NASA's Research Center in Cambridge, Massachusetts.

Kock's interest in breeding orchids began when a collapsible greenhouse took his fancy. It wasn't long before he was growing varieties of unnamed orchids and registering names after his wife and daughter—both named Kathleen—a daughter-in-law, three granddaughters, and a cousin.

Kock directed the Herman Schneider Basic and Applied Science Laboratory from his home in Ann Arbor (after he returned to Bendix from NASA), but he would come to campus each month to advise engineering students on their interdisciplinary research projects. In 1978 he published his advice on creativity in a book, The Creative Engineer. His other seventeen books were scientific, but at the time of his death in 1982, he had begun work on a second book on creativity, Increasing One's Ability to Invent. Although a scientist by profession, Kock attributed his creativity to his two passions: music and chess.

Adapted from ". . . A Legacy of Electronic Marvels" in Cincinnati Horizons, February 1983.

"My memories of college life center around my days of living on campus and the Sunday Bengals football games played at Nippert Stadium. Watching the game from inside one of several buildings surrounding the stadium was a challenge of sorts."
—Jim Ingmire, B.S.Ch.E. '73

its new name—was created not as physical space, but as a source of seed money for college faculty to pursue research in existing laboratories. Winston Kock, an electrical engineering graduate of the Class of 1932, was appointed director.

More help for faculty development came quite unexpectedly in 1974, in the form of a $1,700,000 gift—at that time the largest bequest ever received by the college. The benefactor was Lester C. Morrow, a mechanical engineering graduate of the Class of 1912. He was founder, editor, and publisher of *Factory* magazine and a distinguished alumnus honored at the cornerstone ceremony. Morrow had donated five hundred dollars to the Sesquicentennial Fund for unrestricted uses, and shortly after the celebration he announced his desire to write a bequest to the college in his will. At the dean's suggestion, Morrow designated the money to support the individual professional development of faculty members, over and above the general funds of the university. The fund could also be used for seed money to start approved experimental or publication projects, for classroom development, and for continuing education for faculty.

In 1974 Wandmacher departed from the dean's office and returned to teaching in the Department of Civil and Environmental Engineering. That year he also served as president of the American Society for Engineering Education. He was replaced by Clyde Melvin Adams, Pelton Professor of Materials Engineering from the University of Wisconsin-Madison. More departments expanded during the two years that Adams was dean. The Department of Aerospace Engineering became the Department of Aerospace Engineering and Applied Mechanics, the Department of Mechanical Engineering became the Department of Mechanical and Industrial Engineering, and the Department of Electrical Engineering became the Department of Electrical and Computer Engineering.

In 1975 the deans of the co-op colleges decided to create a Division of Professional Practice separate from the Career Dynamics Center, with the rationale that co-op is an academic program and placement of graduating seniors is a student service. The separation established professional practice as a distinct academic unit, but retained a system to coordinate policies and operations among the co-op colleges.

In the early seventies the Department of Mechanical Engineering began to offer a complete curriculum in industrial engineering, expanded from an industrial management program in which students took two years of

Dean Adams, standing, is speaking at the 1975 Distinguished Alumnus Awards banquet.

Robert Delcamp held major administrative responsibilities in the first official recruiting effort of women and minority students, which began in 1969.

"One of my most memorable faculty members was 'Doc' Price. Doc in his own way was truly a very warm-hearted and nurturing individual. I don't think I'll ever forget the first time one of our class members asked him what the correct answer was, and he commented, 'How should I know—nobody's ever got that far.'"
—Robert J. Ruebsch, B.S.Ch.E. '70

The Department of Electrical Engineering had always been in Swift Hall, until Rhodes Hall was completed in 1970. This is an early 1970s photo of a clean room in the semi-conductor laboratory in Rhodes Hall.

engineering courses and two years of business courses offered jointly from the business and engineering colleges. This new program was formed in 1975 by Dr. Richard Shell, head of the Department of Management in the College of Business Administration, who transferred to the College of Engineering. The following year the UC Board of Trustees approved a change of name from the Department of Mechanical Engineering to the Department of Mechanical and Industrial Engineering. For the next four years, graduates of that program earned degrees in mechanical engineering with a concentration in industrial engineering. Then, in 1981, the Ohio Board of Regents approved the degree of Bachelor of Science in Industrial Engineering.

The mechanical engineering department's Structural Dynamics Research Laboratory, now directed by David Brown, A.E. '61, M.S.A.E. '63, Ph.D. '76, entered a new phase of capability in the late seventies with developments in computer and electronics technologies. The SDRL pioneered the development of multiple degrees of freedom algorithms jointly with a research group at the University of Louvain in Belgium. Their microcomputer-based system allowed for real-time animation of modal vectors for the automotive industry, providing the ability to interpret the physical meaning of vectors. Hardware and software developed by the SDRL became the basis for many commercial systems for experimental structural dynamics.

In 1978 the SDRL offered the first of many intensive short courses in experimental modal analysis as an educational service to industry, and to generate funding for its research and personnel. The income from the courses, together with grants from Hewlett-Packard, has enabled the SDRL to maintain leadership in equipment and research in experimental analysis of the dynamics of structural systems.

On more than one occasion, the Schneider Medal has been awarded to two students from one class. In 1980 the medal was awarded to Robert Ryan (B.S.M.E.), left, and Mary Ell (B.S.C.E.), right. Dean Kirk Valanis is in the center.

Dean Adams resigned in the fall of 1976 to pursue research and teaching, and Associate Dean Robert Delcamp became acting dean of the college. Delcamp had been a chemical engineering faculty member since 1945, had earned his doctorate at UC, and was a research associate in the Tanners' Council Research Laboratory from 1954 to 1962. He remained in that role of acting dean for the next two years.

One of the most significant changes in the university's history came the year after Adams resigned. In 1977 the Ohio Board of Regents and the UC Board of Trustees approved the decision for the university to become a full-fledged state school. Although the University of Cincinnati was no longer a municipal institution, it was now eligible for more state funding. The new state university had a new president that same year when Warren Bennis was succeeded by Henry Winkler, who began his teaching career in 1939 as an instructor in history for the College of Engineering. The following year Dr. Kirk C. Valanis, chair of the Division of Materials Engineering at the University of Iowa, came to UC as the new dean of the College of Engineering.

The Department of Chemical and Nuclear Engineering helped to form a Polymer Research Center in 1979, directed by Professor J.E. Mark. The center was established to coordinate course offerings and research for graduate students, from any department, wishing to specialize in polymers, and to sponsor courses for undergraduates. In 1981 the Department of Mechanical and Industrial Engineering formally established the Ergonomics Research Laboratory,

President Henry Winkler began his teaching career in 1939 as a graduate instructor of history in the College of Engineering.

"Although co-oping allowed us to cover most of our school expenses, we were usually limited to less expensive diversions such as a 70-cent slice of pizza from Papa Dino's and a $2.99 twelve-pack from Joey's Deli."
—Joseph R. Lancauskas, B.S.E.E. '84

A History of Research with the EPA

On July 3, 1975, President Gerald Ford came to Cincinnati to dedicate the National Environmental Research Center. It is the US Environmental Protection Agency's national headquarters for drinking water research, water treatment, and hazardous waste treatment. Its location, on Martin Luther King Drive adjacent to the University of Cincinnati, makes it possible for the Department of Civil and Environmental Engineering to maintain its long-standing cooperative endeavors in environmental engineering with the EPA.

The predecessor of the EPA is the United States Public Health Service (USPHS), which originated in Cincinnati in 1912 as the nation's first water pollution laboratory. Sanitary engineering at the university has its roots in the Tanners' Council Research Laboratory, with research in treating industrial wastes in the tanning industry. The co-op program in sanitary engineering began in 1956 when Dean Wandmacher recruited John David Eye to be the college's first professor in that field. Eye's first undergraduate students were two civil engineering sophomores, John Kittrell and James Morand, both of the Class of 1960, to whom the dean gave permission to take biology and chemical engineering courses in lieu of some applied civil courses. That gesture alone was an innovation, since electives for co-ops were unheard of at the time. In 1958 they shared the first co-op position in sanitary engineering at the USPHS Robert A. Taft Sanitary Engineering Center, the national headquarters for air and water pollution research.

In the early sixties, more students in sanitary engineering were gaining experience through government training grants at the Mill Creek Treatment Plant, a USPHS testing facility next to the city's sewage treatment plant on Gest Street. When the grants were phased out, Louis Laushey, chair of the Department of Civil and Environmental Engineering, and Professor Pasquale Scarpino worked with Andrew Breidenbach of the EPA to initiate a program for co-ops and graduate students to conduct experiments and operate the test facilities.

Center Hill, a research facility contracted to UC by the EPA, was built in 1967 on land which the Procter and Gamble Company transferred to UC in the fifties as a site for a nuclear reactor. The funding for the reactor never materialized, and the land lay dormant until Laushey and Professor Herbert Preul initiated the effort to turn the site into a laboratory for research in solid waste.

In 1970, when the USPHS became the Environmental Protection Agency, it split its research interests at the Taft Center into two separate facilities, one for air and one for water pollution. The EPA retained in Cincinnati all research activity in drinking water, waste water treatment, and liquid and solid hazardous waste treatment at a new site, on Martin Luther King Drive (which was then St. Clair Street) between UC's east and west campuses. To continue the joint research between the neighboring institutions, and to establish a system of shared facilities, Breidenbach, who became the first director of the new EPA center, and UC President Warren Bennis formed a Linking Pins Committee of representatives from UC, the EPA, and the city. The mutual benefits include personnel exchanges, symposia, and shared use of laboratories, libraries, and computers. In 1980, five years after President Ford dedicated the new facility, it was renamed the Andrew W. Breidenbach Environmental Research Center. Since its founding, the center has trained co-op students and hired many UC graduates, in addition to its joint research with the university.

President Gerald Ford came to Cincinnati on July 3, 1975, to dedicate the Environmental Protection Agency's National Research Center. From left to right are Andrew Breidenbach, who was the center's first director from 1975 to 1978, UC President Warren Bennis, Reverend Venchal Booth of the UC Board of Directors, and President Ford. In 1980 the center was renamed the Andrew W. Breidenbach Environmental Research Center, in his memory.

directed by Dr. Anil Mital. That laboratory participated in founding the *Journal of Industrial Ergonomics*.

General Electric and the College of Engineering expanded their joint program of advanced training for professional engineers when they introduced the Advanced Course in Manufacturing (ACM) in 1981. Like the Advanced Course in Engineering, which is exclusively for aerospace engineers, the ACM involves UC faculty and industrial experts teaching courses in manufacturing problems, leading to a graduate degree in mechanical, industrial, electrical, computer, or metallurgical engineering, as well as materials science and interdisciplinary studies.

Two research centers, one in robotics and one in membrane technology, were established in 1983. The university established the Center for Robotics Research in the Department of Mechanical and Industrial Engineering, with Ernest L. Hall as director. The purpose of the center is to perform interdisciplinary research in intelligent robotics systems, building on the college's previous work in mechanical design, machine intelligence, advanced manufacturing systems, and human engineering. The Membrane Technology Center, directed by Sun-Tak Hwang, was established by a grant from Standard Oil of Ohio for research in membrane separation processes for medical and industrial applications. The research is centered on developing more permeable and more selective membranes. Applications include removing salt from salt water, cleansing blood through hemodialysis, oxygen enrichment devices for lung disease patients, as well as industrial applications.

In 1984 Dean Valanis resigned and President Winkler retired. To immediately replace Valanis, Louis M. Laushey accepted an appointment as interim dean, pending the appointment of a permanent dean. Laushey was recruited to the college by Dean Wandmacher in 1958, as his successor to the William Thoms Chair of Civil Engineering. Laushey chaired the Department of Civil Engineering for twenty consecutive years, until 1978. During those years, he served as the faculty representative to the University Board of Directors from 1968 to 1969, and served on the Faculty Advisory Committee to the Chancellor of the Ohio Board of Regents between 1968 and 1985. He was vice-chairman and then chairman of that committee from 1979 to 1983.

Joseph Steger, senior vice-president and provost of the University of Cincinnati since 1982, succeeded Henry Winkler. He was inaugurated on November 30, 1984. Norman R. Baker, a professor of quantitative analysis in the College of Business Administration since 1974, became the new senior vice-president and provost. Baker earned a doctorate in industrial engineering and management science from Northwestern University in 1965, and had taught industrial engineering at Purdue

"Towards the end of spring quarter 1978 the amateur radio club, W8YX, acquired a rather large amateur antenna which was to be mounted on top of one of the towers atop Swift Hall. It took us all day to get the antenna just to the top of the tower. Dusk was fast approaching and I didn't want to leave this large antenna unattached at the top of the tower overnight. Then I got an idea. What if the lights of Nippert Stadium could be turned on? I thought this might light up the skyline enough so that we could continue with the antenna installation. Dr. Carl Osterbrock had given out his home telephone number to those of us taking his field theory course so that if we had any questions about homework problems or just questions in general, we could call him. I called him and he said he would see what he could do. It wasn't twenty minutes later that the lights of Nippert Stadium began to come to life. Everybody on the roof of Swift Hall looked at each other in amazement. Boy! Was that service! We worked until about 1 a.m. attaching the antenna. I don't know what the light bill was for six hours of lighting Nippert, but I doubt that we students could have pulled it off by ourselves."
—Brad Smithwick, B.S.E.E. '81

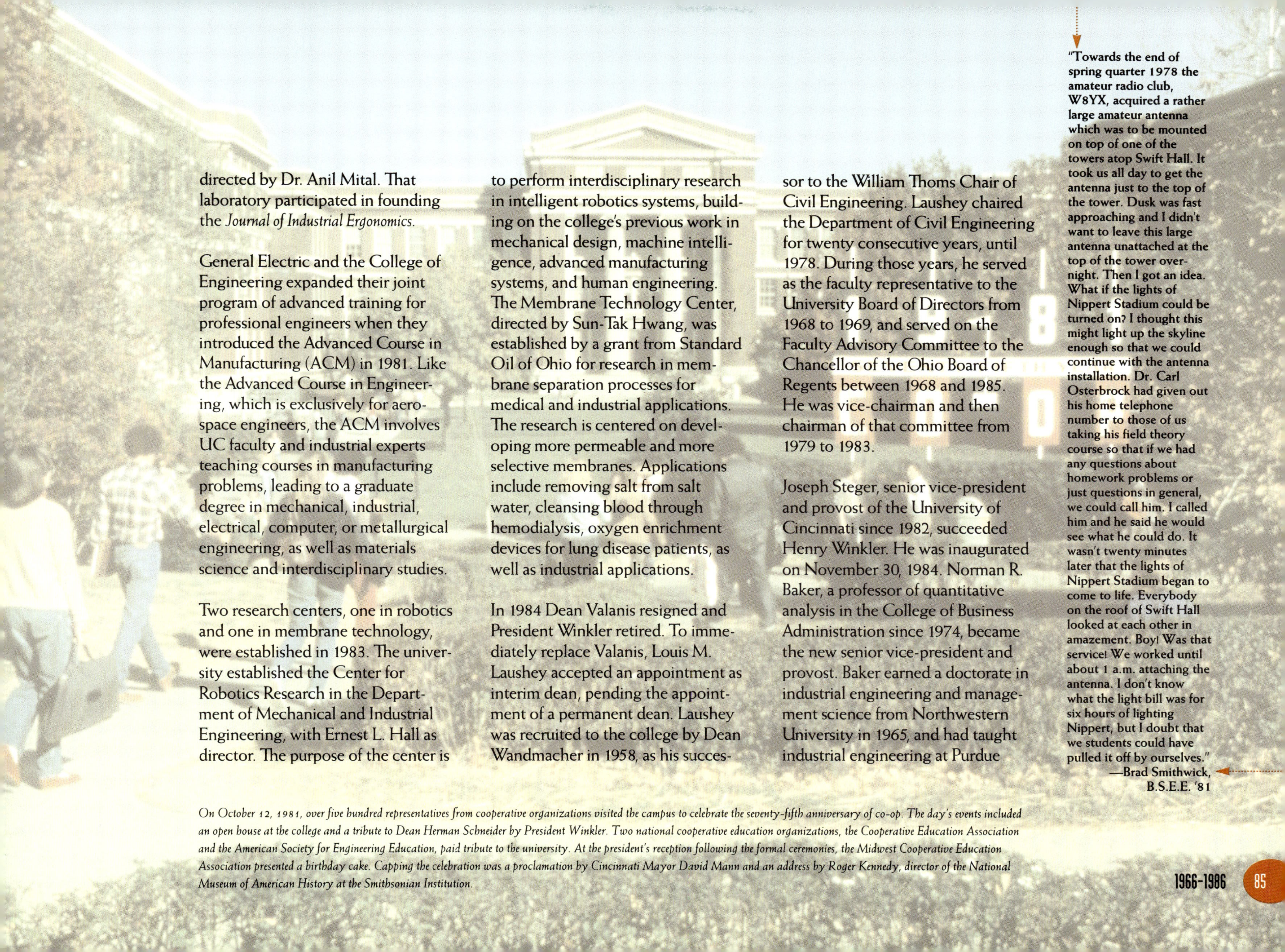

On October 12, 1981, over five hundred representatives from cooperative organizations visited the campus to celebrate the seventy-fifth anniversary of co-op. The day's events included an open house at the college and a tribute to Dean Herman Schneider by President Winkler. Two national cooperative education organizations, the Cooperative Education Association and the American Society for Engineering Education, paid tribute to the university. At the president's reception following the formal ceremonies, the Midwest Cooperative Education Association presented a birthday cake. Capping the celebration was a proclamation by Cincinnati Mayor David Mann and an address by Roger Kennedy, director of the National Museum of American History at the Smithsonian Institution.

Louis Lausbey neither expected nor asked for the position of interim dean. But he accepted President Steger's request, and pledged to give his utmost to the college.

University and the Georgia Institute of Technology before coming to UC.

Industrial engineering on the UC campus experienced major growth in 1985 when the Department of Mechanical and Industrial Engineering received a long-term training grant from the National Institute of Occupational Safety and Health to offer graduate industrial engineering degrees in occupational safety. In its safety engineering research, the department has worked closely with the Department of Environmental Health in UC's College of Medicine.

Also in 1985 came the fruition of a three-year joint effort among industry, local and state government, and the university to plan a major research institute aimed at improving profitability and competitiveness of manufacturing industries in the area. Planning for the Institute of Advanced Manufacturing Science (IAMS) began in 1982. Founding partners were the University of Cincinnati, General Electric Aircraft Group, Cincinnati Milacron, the Kroger Company, the Procter and Gamble Company, and Cincinnati Gas & Electric Company.

For the project, the Greater Cincinnati Chamber of Commerce and the City of Cincinnati commissioned a study of how to upgrade the area's industrial base and improve economic growth through expanded business development. The study found that the area economy was mature, well-diversified, and stable, but had experienced little real growth for the last two decades. The recommended strategy for developing the region through advanced technology was first to integrate advanced manufacturing systems into the existing industrial base and then to develop an electronics assembly industry. Battelle Institute's laboratories in Columbus surveyed local manufacturing firms and identified areas of research with the most potential for IAMS. The areas included robotics, computer-aided design and manufacturing, manufacturing automation, computer control, and improved productivity.

In 1983 the City of Cincinnati designated land for a research park site, and the next year the state approved $8 million to construct a facility; the State of Ohio Thomas Edison Partnership Program provided $4.1 million in initiation funds. In November 1985 Charles F. Carter, Jr., from Cincinnati Milacron, became the first full-time executive director of

"I began co-oping in 1983 and at that time the job market for mechanical engineers was not too good. Finding a co-op job was tough. However, by graduation the job market was booming. The atmosphere of school was more economically driven than political. In fact, one time G. Gordon Liddy spoke on campus and a group of demonstrators were giving him a hard time. Some friends and I got fed up with their chanting of slogans and made them be quiet. The crowd applauded our efforts."
—Pete Riley, B.S.M.E. '85

Throughout its history, the college has benefitted from the generosity of hundreds of alumni. One such benefactor was Joseph Dave (C.E. '23), founder of the Dave Steel Corporation, who established a scholarship fund in 1975. The design and construction faculty of the civil engineering department joined the Daves in awarding the first scholarship. From left to right are Andrew Bodocsi; Harvey Ingram of the UC Development Office; Philip Meyers of the UC Board of Trustees; Bernard Dave, president of the Dave Steel Corporation; Michael Basehart; Louis Lausbey; Issam Minkarah; Joseph Dave, Joseph Steinbicker (C.E. '76), first recipient of the scholarship; John Cook; Boyd Ringo; and Cornelius Wandmacher.

THE AFGHANISTAN PROGRAM

In 1963 the College of Engineering joined a consortium of nine other American universities and the US Agency for International Development (USAID) to establish an American-supported College of Engineering at Kabul University in Afghanistan. As a member of the USAID steering committee, Dean Wandmacher visited Afghanistan twice. One unplanned outcome was for the dean to recruit Robert K. Michael, then involved in establishing the laboratories in Kabul, to do the same for the new Rhodes Hall at UC.

But the UC faculty member who was the most involved in the program was civil engineering professor James McDonough. Wandmacher first sent McDonough to Kabul to teach basic civil engineering courses in 1969. McDonough stayed there until 1971, when his brother John, an instructor in the Department of Engineering Analysis back at UC, took his place for another two years.

Throughout the 1970s this undergraduate program produced dedicated engineers, some of whom became technical leaders in Afghanistan. Many came to the United States for graduate studies at the participating universities. Fifteen young Afghan scholars who were prospective faculty at Kabul enrolled in various departments in the College of Engineering at UC.

But beginning in 1978, many Afghan faculty and engineers fled to the US and other countries for a different reason. The Russian Army invaded Afghanistan that year, and a communist revolution seemed imminent. Kabul University, with its close American affiliation, was an obvious target for communist rebels.

During the war, the engineering classrooms were completely looted and most faculty and students fled the country; when the school was closed in 1979, the faculty roster of one hundred had by then diminished to twenty. By the time the war ended, ten years later, five million refugees had fled their villages and most roads and irrigation systems were damaged and even destroyed.

That year James McDonough went to Pakistan, and with Professor Edward Friedman of the Stevens Institute of Technology—one of the ten original consortium universities—he wrote a proposal for Construction Related Training for Afghanistan Rehabilitation and Reconstruction. In 1990 the USAID Office of Afghan Affairs initiated the program in Peshawar, Pakistan, where many of the Afghan engineers lived during the war. The program was established to retrain Afghan engineers through short intensive courses to bring their knowledge up-to-date.

Engineering students whose final year of coursework was interrupted by the war can obtain certification equivalent to a bachelor of science degree in civil engineering, so they can participate in rebuilding their country's infrastructure. Dr. A. Wahed Hassani, coordinator of the program, recruited faculty from six American universities. In the summer of 1991 McDonough taught courses in reinforced concrete and construction management to twenty-eight students, including Dr. Hassani. Since the midday temperature in Peshawar reached 110°, his classes were usually held from 3:00 to 5:00 in the afternoon. "We developed all new lecture material from the original curriculum," said McDonough. "It was some of the hardest work I had ever done."

Twenty-eight Afghan engineers and engineering students attended James McDonough's civil engineering courses, as part of the Construction Related Training for Afghanistan Rehabilitation and Reconstruction in Peshawar, Pakistan, in the summer of 1991. McDonough is seventh from the left in the first row.

"My engineering education has served me well. I always believed that school taught you how to think, and gave you tools and building blocks from which experience became expertise. The engineering education taught me how to reduce large problems and situations into smaller, more manageable pieces. This is the cornerstone of essential business practice."
—Christopher John Whitaker, B.S.A.E. '82

IAMS, and in December ground was broken for a 64,000—square foot building in Bond Hill, near Interstate 75 and Paddock Road.

By the middle of the 1980s, research and teaching in engineering at the university level had become inseparable. For the rest of the decade, the College of Engineering continued to strengthen its ties between those two areas, as the process of becoming a premier research institution accelerated.

President Winkler, right, presents the George B. Barbour Award to chemical engineering professor Robert Delcamp, center, at the June 1982 commencement exercises in Nippert Stadium. On the left is mechanical engineering professor Ronald Huston, who was acting provost at the time. Since 1969 the Barbour Award has recognized annually one professor, out of all UC faculty, who has contributed the most to further good student-faculty relations. Two other engineering professors have won the award: Richard Shell, mechanical and industrial engineering, in 1980, and Boyd Ringo, civil engineering, in 1986. George Barbour, a professor of archeology, was dean of the College of Arts and Sciences from 1938 to 1958.

CHAPTER 6
1986-1992

A Premier Teaching and Research Institution

In his research, electrical engineering professor H. Thurman Henderson developed a new family of gated deep-impurity double injection semiconductor devices. In the late 1980s the State of Ohio awarded the Department of Electrical and Computer Engineering all three of Ohio's Selective Excellence Awards.

FUNDED RESEARCH IN THE COLLEGE had its origins in the Basic Science Research Laboratory of the 1920s and 1930s. In later decades other research programs emerged to establish the college's focus in graduate education. Therefore, it was only natural in the 1980s, when President Steger and Provost Norman Baker began turning the University of Cincinnati into a premier research institution, for the College of Engineering to be the first to follow suit. The president's interest in graduate education and corporate relations for the university stemmed from his own background in business as director of organizational development and human resources at Colt Industries, and before that, as dean of the School of Management at Rensselaer Polytechnic Institute. Steger's background, matched with Baker's experience in industrial engineering, fostered the right climate for an unprecedented period of growth in graduate education and research.

Within the university, the College of Engineering was a logical place to

The background of President Joseph Steger, right, in business and management, and the background of Provost Norman Baker, left, in industrial engineering, fostered the right climate in the 1980s for turning the university into a premier research institution.

start. After eighty years of cooperative education, the college had established hundreds of contacts in industry, and its six departments were already engaged in research through their graduate programs. The timing was appropriate as well, since the college was in the midst of a national search for a dean and was therefore preparing for a change in leadership.

In 1986 Constantine Papadakis filled the position. He earned a master's degree in civil engineering from UC in 1970, and a doctorate from the University of Michigan in 1973. He was returning to UC's College of Engineering from Colorado State University, where he was head of the Department of Civil Engineering. Before his tenure at Colorado State, he had been assistant chief engineer for Bechtel Power Corporation, vice-president of STS Consultants, Ltd., and then vice-president of Tetra Tech, a Honeywell subsidiary. His appointment marked the first time someone with significantly more experience in industry than academia would occupy the dean's office. But the search committee felt that his experience in both business and education were the right combination for the new directions the college would take.

As dean, Papadakis was responsible for expanding the focus of the College of Engineering from its strong tradition of undergraduate education to include a parallel emphasis in graduate studies and research. This approach enriched the college's undergraduate programs even further; professors would stay at the forefront of rapidly changing technologies, and students could witness discovery and learn that they have the ability to be a part of the process of technological innovation.

Balancing the demands of providing this type of undergraduate education with cultivating research opportunities called for another key administrative position, that of an associate dean for academic affairs, which James McDonough, Ph.D. '68, assumed in January 1987. McDonough had been head of the Department of Civil and Environmental Engineering since 1979. While Papadakis focused on building the research initiative, McDonough chaired the college's Curriculum Committee, and became responsible for student recruitment, enrollment, retention, and the quality of teaching.

Papadakis mobilized the college faculty by calling for all departments to formulate their own goals, objectives, and tactical plans. Together,

Constantine Papadakis earned his master's in civil engineering from the University of Cincinnati in 1970. When he returned to UC in 1986, he became the first dean with significantly more experience in industry than academia. This qualification uniquely suited him to lead the college in a new direction toward prominent research.

In 1991 the first minority graduates of the college reunited at UC. From left to right the "pioneers" are: Lamar Cox, E.E. '56; Henry Brown, Ch.E. '55; Clark Beck, M.E. '55, M.S.A.E. '69; Leo Wilcox, C.E. '57; Albert Andrews, A.E. '57; James Wade, A.E. '58; Marvin Foster, E.E. '56; and Harold Crosby, E.E. '56.

the faculty and administration developed a college-wide strategic plan, to concentrate on six over-arching goals. Two of the goals would build on existing strengths: to attract and nurture quality faculty and students, and to encourage the faculty to develop an enriched educational experience for all students.

Four additional goals brought the college into new arenas. These goals aimed to develop a multi-cultural learning environment, establish prestigious research activity, create and maintain the infrastructure necessary for quality instruction and research, and develop a comprehensive fundraising program to provide the means of achieving these goals.

Building the educational infrastructure called for integrating computing into the college's everyday life through a three-phase process. In the first phase, all computer hardware in the college was networked to the university's mainframes and to the Ohio Supercomputer Center in Columbus. The network's design and construction were spearheaded in 1986 by mechanical engineering professor David Brown, the director of the Structural Dynamics Research Laboratory.

The second phase provided a personal computer for every faculty member, and workstations for those using advanced computing in their work. Several departments provided clusters of workstations for their students as well. Harold Carter, a professor in the Department of Electrical and Computer Engineering, contributed significantly to this second phase. When Jack Krebs was hired as director of college computing in 1989, he began the third phase, integrating computing throughout the curriculum. Under his direction, seven computer laboratories opened for students to use twenty-four hours a day.

New programs to recruit and retain minority students were established by Edward Prather, who became the assistant dean for minority affairs in engineering in 1988. He came from the Stevens Institute of Technology, where he directed the Technical Enrichment Program. With the help of Sammie Brooks, Jr., an engineer on loan from the IBM Corporation, Prather recruited an advisory council and secured mentors for African-American students.

He also developed a host of college preparatory programs, including the Summer Enrichment Program in mathematics and science for students in grades six through eleven, and the Summer Institute, for the same age group, to develop conceptual understanding and problem-solving skills. Prather's office initiated the Bridge Program to help incoming

"At least once a quarter, without fail, some poor soul tries to get to the eighth floor via the center stairs of Baldwin. (The stairs are permanently closed off because of the expansion of the library). Even the Dean is guilty."
–Beth Brock, M.E. '93

CO-OP LIFE HAS NEVER BEEN ALL WORK

In 1912 twenty members of the University of Cincinnati Bearcats were co-ops.

"I really loved it when Bob Huggins was brought in as basketball coach and turned a rather weak team into a national power. With last year's Final Four trip and this year's success, I've felt much like the Bearcat fans of the great basketball teams of the past."
–Dennis M. (Mike) Miles, M.E. '93

Spending half of their college careers off campus has made it challenging for co-ops to get the most out of college life. Student organizations have always been one way to bridge that gap between their freshman and senior years, which are the only times when engineering students take classes all year long.

It was in athletics that the students made one of their first marks outside formal education. The university organized its football team in 1885, and by 1906 the student engineers comprised the largest contingent on the Bearcats. Having co-ops on the team required that football practice be held at night, since they couldn't leave work for practice during the day. Some rudimentary form of illuminating Carson Field was tried as early as 1906, but it wasn't until 1909 that the Athletic Council paid for arc lights mounted on wooden poles around the field. The football was painted white and night practice was born—just at the right time. By 1912 twenty of the twenty-eight squad members were co-ops.

In November 1921 electrical engineering student Jack B. Silverman designed a new system of incandescent lights for the gridiron. The area between goalposts was lighted to a width of sixty yards and a height of forty feet. Silverman wrote his 1923 senior thesis in part on the project, by which time he had enlarged his lighting system (see Silverman's reminiscence on page 37). The team graduated from night practices to night games in 1923,

playing their first against Kentucky Wesleyan. (The Bearcats won, 17-0.)

In 1959 co-op athletics were in a different spotlight, when the NCAA put UC on probation for giving athletic scholarships to co-ops, who were not in the classroom full-time. In a harsh letter to the NCAA, President Langsam accused the organization of "publicly stigmatizing the university," instituting a "glaring injustice," and forcing students in financial need to choose one form of education over another. The NCAA lifted the probation.

Tau Beta Pi, the national engineering honor society, installed the UC College of Engineering as the Ohio Beta Chapter in 1915. The society held its national convention at UC in 1920, 1938, 1961, and 1975. It was at the 1961 convention that Dean Justice accepted the Bent monument from the association, as a tribute to Dean Schneider.

The traditional competitions among the departments in the college during E-Week include a tug-of-war.

Each quarter the pledges of Tau Beta Pi are required to polish the Bent monument.

One of the oldest organizations, the Engineering Tribunal, was founded in 1907 as a governing body, assuming a judiciary role in conduct and ethics, and helping to plan and direct the other extracurricular organizations. The tribunal also acted in an ombudsman capacity, mediating complaints from both faculty and students.

In 1986 the tribunal introduced UC to National Engineers Week, sponsored by the National Society of Professional Engineers, and gave Dean Schneider his most informal tribute ever—a birthday party. On Monday, February 18, the engineering students wore "Thanks, Herman!" butttons, served birthday cake, and invited alumni who knew the dean to speak at the party.

Each year during E-Week, traditionally held in February, students from all departments compete in trivia competition, a blood drive, and tug of war. The most popular event of E-Week has come to be the egg drop contest, where students design devices to cushion a raw egg from a twenty-foot drop.

Sixty-eight organizations have existed throughout the history of the College of Engineering. See the Appendix for a complete lists of their names and founding years.

"I didn't really find out about E-Week until my senior year because my co-op quarters were summer and winter. E-Week was a lot of fun to participate in and watch. My favorite event to watch was the egg drop contest off of the Zimmer Plaza. People came up with all kinds of ways to protect the egg from being crushed on impact. There were parachutes, foam, and even Jell-O among the design entries. Whoever was able to get their device the closest to the target with the egg intact won the event. E-Week was a chance for seniors who had worked on opposite co-op sections to get to know each other better outside the classroom. You don't realize how many people are actually in your class until you return for your last year and meet all these people you haven't seen in three years."

—Michael Buhr, B.S.E.E. '89

Farhad Radpour, a professor of electrical engineering, conducts research in long wavelength infrared detectors, using multiquantum wells and superlattice structures. He synthesizes structures of semiconductor materials, atomic layer by atomic layer, that do not exist in nature. He is standing at a metal organic chemical vapor deposition (MOCVD) facility, which is used to build these structures.

freshmen adjust to college-level courses, and the Cooperative Learning Program, to encourage peer interaction among students.

In August 1988 Papadakis formed the College Advisory Council with the help of William Dickhoner, M.E. '50, who assisted him in recruiting executives across the nation, many of whom had ties to the college as alumni, through co-op, or in joint research and development. After forming the council, the seventeen charter members defined the council's missions to address the college's strategic issues, and to identify how the technical, research, and business trends of industry would influence the college's future directions. The council's semiannual meetings thus claimed a permanent and valuable place on the college calendar.

"It's a tradition to walk through the circle outside of Langsam Library before a big test. The circle is thought to bring good luck when necessary, however it is never used until exam week."
–Michael A. Strader, M.E. '93

The Ohio Board of Regents Selective Excellence Program, first implemented by Governor Richard Celeste's administration, became an early source of funding for the college's strategic goals. All departments in the college were recognized by either the Academic Challenge Award, the Eminent Scholar Award, or the Program Excellence Award. Five of the departments were noted for existing excellence and proven strengths with the Academic Challenge Award, which supplemented each program's budget by ten percent. In total, the Academic Challenge Awards financed positions for nine new faculty members in those five departments.

Andrew Steckl, Ohio Eminent Scholar and Gieringer Professor of Solid State Microelectronics inserts a silicon wafer, covered with a silicon carbide film, into a reactive ion etching system.

Through statewide competition, four departments in the College of Engineering—more than any other single college in the state—won Ohio Eminent Scholar Awards. These awards established endowments for Ohio Eminent Scholar Chairs by providing each department with $500,000, to be matched by additional million-dollar gifts. The Ohio Eminent Scholar Awards were granted to the College of Engineering in Solid State Microelectronics, Advanced Manufacturing, Experimental Fluid Mechanics, and Membrane Technology.

The Department of Electrical and Computer Engineering became the first at UC to win all three Selective Excellence Awards. The department had been prospering steadily for years under the direction of Thurman Henderson, and then under Vikram Kapoor, who succeeded Henderson in 1986. As department head, Kapoor upgraded the department's solid state research laboratories, and, with the help of Professor William

"During the 1991-92 Spring Quarter, the industrial engineering students were overloaded with computer work. We were all too familiar with the problems of the engineering computing facilities. As a group we compiled a letter and a list of suggested improvements. Being diplomatic, we gathered petitioners' names. The results were impressive. Jack Krebs (Director of College Computing) responded with action."
–Kyle R. Sasse, I.E. '94

Middendorf, established a baccalaureate program in computer engineering, which was approved by the Accreditation Board for Engineering and Technology in 1987. By 1989 the department's faculty doubled from fifteen to thirty, and research endeavors expanded for the first time into signal processing, automatic controls, computer system design, and applied systems.

One of the faculty recruits was Andrew Steckl, who came from the Rensselaer Polytechnic Institute in 1988 to be the department's Ohio Eminent Scholar. He was hired after a gift of $1 million from Carl Gieringer, E.E. '32, M.S.E.E. '34, D.Sc. '76, chairman of the Ohmart Corporation, supplemented the state award. In occupying the Carl K. Gieringer Ohio Eminent Scholar Chair in Solid State Microelectronics, Steckl became the head of the department's Nanoelectronics Laboratory, conducting research in nanostructures and silicon carbide materials for high temperature electronics. The Department of Electrical and Computer Engineering also won the Ohio Board of Regents Program Excellence Award of $200,000 for laboratory equipment.

The Department of Aerospace Engineering and Engineering Mechanics, the second oldest in the nation—also expanded in response to several opportunities. Throughout the 1980s the Gourman Report's annual evaluation of academic programs at

In the Center for Excellence in Propulsion Technology, Awatef Hamed, pictured here at the nozzle of a rocket engine, develops ways to improve the performances and increase the lives of propulsion systems in both airplanes and rockets. The center features a unique high-temperature gas-particle tunnel, to which aircraft engine manufacturers from across the nation come to test their components.

As a graduate student, David Carson, B.S.C.E. '85, M.S.C.E. '88, monitored permeability tests of grouts for containment of hazardous wastes.

COLLEGE AEROSPACE ENGINEERS AT NASA

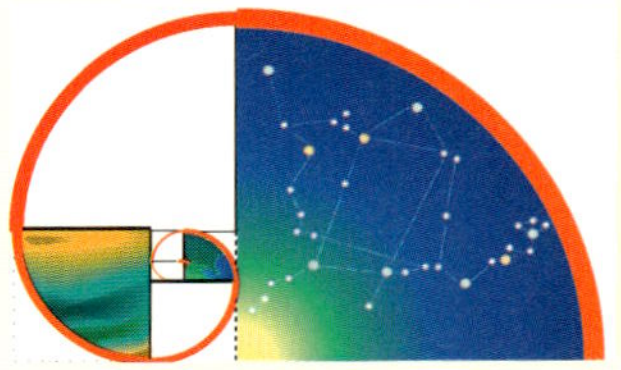

After the space shuttle Challenger exploded in 1985, NASA made substantial improvements in its safety programs, including increased research funding. In April 1988, through a competitive grant program, NASA awarded the College of Engineering $5.1 million to establish the Health Monitoring Technology Center for Space Propulsion Systems, one of nine joint NASA/university research centers. Through research and student training, the center provides NASA and industry with resources for developing safe and reliable space propulsion systems for the next generation of reusable space vehicles. In 1992 NASA renewed the grant and increased its budget in recognition of the college's successful research program.

The research conducted in the UC center is multidisciplinary; faculty and students participate in the long-term research goal of determining procedures for reconfiguring propulsion systems and for operating strategies to respond to detached or predicted deterioration, in order to accomplish mission objectives and insure safety. Reliable online monitoring of reusable space propulsion systems requires the inboard real time diagnosis and prognosis of deterioration and its effects on performance. The center conducts analytical and experimental research to provide a comprehensive educational program that emphasizes all technological areas required to monitor space propulsion systems. They include interplanetary systems, space station transports, orbital transfer systems, and reusable launch systems. General Electric Aircraft Engines Company is UC's charter partner in the center.

The NASA center fostered an environment for more extensive research in the aerospace industry. In the autumn of 1988 the Ohio Board of Regents appointed Dean Papadakis chair of the committee of engineering professors who organized the Ohio Aerospace Institute (OAI), an idea originated by John Klineberg, director of the NASA Lewis Research Center in Cleveland. The OAI is a nine-university consortium at NASA Lewis for aerospace research, graduate and continuing education, and industrial interrelations. The OAI was incorporated in May 1989, with Dean Papadakis as interim president and UC professor Donald Stouffer as interim Director of Research and Academic Programs. The OAI augments the curricula of the nine universities with visiting faculty and industrial scientists.

In its first year the OAI was awarded funding through the NASA Space Grant College and Fellowship Program. The aim of the funding, which lasts five years, is to broaden the bases of universities' contributions to the aerospace industry and to recruit and educate professionals in aerospace science and technology, by providing undergraduate scholarships and internships, with an emphasis on minority recruiting. The institute also awards doctoral fellowships to students studying at the consortium universities, who spend residencies at the OAI for dissertation research.

These programs are at the forefront of aerospace engineering education. Through its OAI partnerships, the College of Engineering has enhanced its own prestige, which was recognized nationally throughout the 1980s. In fact, the Gourman Report consistently ranked UC's graduate program in aerospace engineering among the top ten in the nation.

At the April 28, 1988, inauguration of the NASA Health Monitoring Technology Center for Space Propulsion Systems at UC are (from left to right): Stanley Rubin, director of the center, Donald Stouffer, author of the winning proposal to NASA for the center and acting head of the Department of Aerospace Engineering and Engineering Mechanics, and Dean Papadakis.

"In my freshman year, there was nothing that I dreaded more than physics tests. All engineering disciplines of the freshman class took the tests in Zimmer auditorium at 5:00 p.m., usually on a Friday. Afternoon tests allowed more time for last minute study, but also left more time for anxiety to build up. The tension peaked at 5:00 when the tests were handed out."
–Thomas G. Umberg, M.E. '93

American colleges and universities ranked UC's aerospace program tenth in the nation. In 1988 members of this department embarked on two major initiatives: the NASA Health Monitoring Technology Center for Space Propulsion Systems and the Ohio Aerospace Institute.

Besides being instrumental in the success of both ventures, Donald Stouffer, the acting aerospace department head in 1987 and 1988, compiled a winning proposal for an Ohio Eminent Scholar in Experimental Fluid Mechanics. When George Simitses was recruited from the Georgia Institute of Technology to become the new department head in 1989, he increased graduate enrollments, added faculty, and upgraded the laboratory facilities. A gift from Percival Roberts, retired vice-president of Towsley Trucks, and his wife, Mary Roberts, B.A. '34, provided for computer laboratories for the department's graduate programs.

The Department of Civil and Environmental Engineering also recruited new faculty and expanded the scope of its traditional programs. In 1988 Paul Bishop came to UC from the University of New Hampshire, to succeed James McDonough as department head. Shortly thereafter he expanded the focus of the environmental engineering curriculum to encompass air pollution, water pollution, solid waste, and especially hazardous waste.

The 1989 Gourman Report nationally recognized this department when, considering its outstanding achievements in air and water pollution and solid and hazardous waste management, it ranked the environmental engineering program number two in the country, second only to the California Institute of Technology.

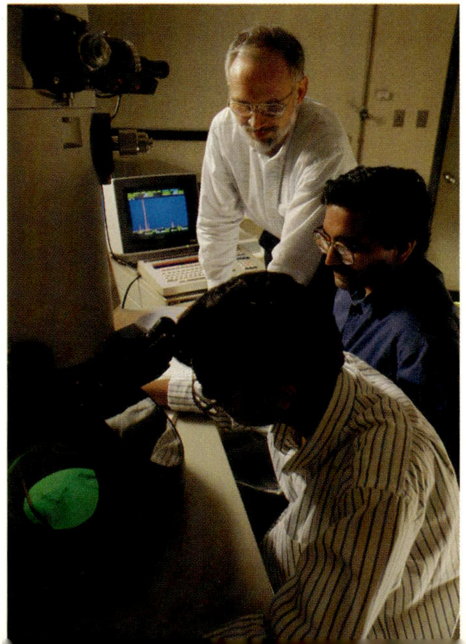

Materials science professors Kenneth Challenger and Vijay Vasudevan, top and middle, conduct research with graduate student Mukul Kumar, bottom, using an anaytical transmission electron microscope.

Professor Emin Aktan joined the civil engineering faculty in 1988, emphasizing structural dynamics and earthquake engineering for graduate studies and research. In 1991, when UC demolished Sander Hall, a twenty-story concrete frame dormitory building, Aktan and his team of graduate students, with visiting researchers from across the nation, installed sensors in the building to monitor its behavior during demolition.

Around this time several college faculty assisted the Smale Commission, formed by John Smale, then chairman of the board at the Procter and Gamble Company, in assessing the condition of Cincinnati's infrastructure. In response to the commission's report of a deteriorating infrastructure, UC established the Ohio Infrastructure Institute, which Dean Papadakis was planning since in 1987. Every engineering college in the state joined this consortium, to

When UC's Sander Hall was imploded on June 23, 1991, an earthquake research team from the college used the unique opportunity to observe and learn from the crumbling structure.

Mechanical engineering graduate student Nancy Hall calibrates an accelerometer during a modal test of the racecar driven by John Andretti in the Indianapolis 500.

develop and test state-of-the-art techniques and materials, to maintain and rehabilitate public works, and ultimately to save taxpayers money. In 1992 the University of Cincinnati, Ohio University, and Case Western Reserve University began to study the deterioration of steel bridges on Ohio's interstate highways. At the same time, The Ohio State University, the University of Akron, and Cleveland State University were planning the construction of a pedestrian bridge in Akron, made of fiber-reinforced polymer composites.

The Department of Materials Science and Metallurgical Engineering also responded to society's need for novel materials by expanding its focus toward a curriculum encompassing metals, ceramics, and polymers. To reflect the shift, the department changed its name to Materials Science and Engineering in 1985. In 1987 the Ohio Board of Regents approved the curriculum, the first of its kind in the state. Following the death of John Moteff, the department's long-time head, the search began for a new leader to manage the significant changes soon to take place.

Kenneth Challenger, Ph.D. '73, was recruited from the Naval Postgraduate School in 1988 to lead the Department of Materials Science and Engineering in an era of unprecedented growth during which the department's faculty roster grew from seven in 1987 to fifteen by 1992. The number of graduate students rose from fifteen to sixty-eight. Challenger also significantly upgraded the department's research arsenal when he convinced Armco and the General Electric Company, two long-time supporters of the department, to provide for a sophisticated scanning electronic microscope in the Kruse Microprobe Laboratory. In 1989 that laboratory's name was changed to the John Moteff Electron Microscopy Laboratory, in memory of the late professor.

In 1990, when Challenger was appointed associate dean for international engineering, he began a college-wide program to provide students with overseas co-op experience in Germany and Japan. Relva Buchanan was then recruited from the University of Illinois at Urbana, where he chaired the program in ceramics, to be the new department head at UC.

Engineering professor Ali Behbahani conducts a Nuclear Reactor Laboratory. Students Douglas C. Lawrence (N.E. '90), David E. Neidich (N.E. '90), and Christopher R. Garnett (N.E. '90), observe as Paul A. Harden (N.E. '90), operates the boiling water reactor simulator.

"One of the most memorable experiences of our college days was the annual ASCE concrete canoe competition. Every year, colleges from Ohio, Kentucky, and Indiana traveled to A.J. Jolly Park in Campbell County, Kentucky with their concrete canoe in tow. Students camped out the evening before the race and enjoyed food, beverages, and some serious euchre games. Race day included various categories of canoe races, a concrete frisbee competition, volleyball matches, and more food and beverages. This event provided the opportunity to meet new people and get ideas to improve on the design for future concrete canoes."
–Sheri Rhodes, C.E. '92 and Michele Landwehr, C.E. '91, M.S. '93

DESIGNING SOLUTIONS FOR REAL PROBLEMS

Since 1990 mechanical, industrial, and nuclear engineering seniors have built on their cooperative experiences through the Design Clinic. In this internship program, the students act as consultants as they solve real design problems of companies in Ohio, Kentucky, and Indiana. In 1991, the program's second year, 118 students worked on 36 projects for 29 companies.

Mechanical engineering professor Lee Harrisberger introduced the Design Clinic concept to the college, after directing a similar but much smaller program at the University of Alabama for ten years. Professor Max Brown was appointed director of the UC clinic, and Elizabeth Seidman was hired as assistant director.

Although the students are not paid, the program provides them with professional experience while offering companies a valuable resource. For a nominal fee, the clinic makes available to each company a team of seniors who collectively devote about seven hundred fifty hours to solving a problem the company provides.

To qualify, a company must have an operational problem requiring engineering consultants. One team helped the Springer School in Cincinnati design a mathematics computer program for learning-disabled children. Before the Design Clinic stepped in, the children were not learning from the program; they were instead entering numbers at random. Rather than rewriting the program, as they originally anticipated, the engineering students researched learning disabilities and discovered that the problem lay not in the program, but in its instructions. They replaced the written instructions with an audio tape and worksheets.

In 1992 Hillshire Farm & Kahn's redesigned part of its meat packing operations and invested $400,000 in new equipment, based on the clinic's recommendation to reduce the risk of injuries at the plant.

The clinic begins in fall quarter with an introductory course to develop the students' professional skills. Winter and spring quarters focus on corporate problem-solving. After an initial visit to their assigned companies, the student teams work in their own space on campus, and send progress reports regularly. When they complete their projects, the teams give formal presentations of their design results.

Three additional programs have grown out of the Design Clinic at UC. The Graduate Research Internship Program offers advanced consulting. The Cost Saving Clinic assists companies with problems of a financial nature, by providing teams of engineering and accounting seniors. The Venture Clinic makes available engineering and business seniors to advise entrepreneurs in developing their products to a marketable stage. All of the programs have the same objective: to provide realistic experience in the practice of problem-solving in engineering. Students gain experience with challenging design problems, and companies receive consulting services for a nominal cost.

During the 1990-91 school year, Jeff Cordray, B.S.I.E. '91, was a Design Clinic consultant to the Clermont County facility of U.S. Precision Lens. His team analyzed and improved parts of the company's manufacturing equipment, to increase operation efficiency.

Sun-Tak Hwang, head of the department of Chemical Engineering, and gradute student Kwang-Rae Lee work on the separation of chemical components in the Membrane Technology Research Laboratory. After Lee completed his Ph.D. in chemical engineering, he was appointed to the faculty of Kang Won University in Korea. He returned to UC in 1990 as a visiting scholar.

Like Materials Science and Engineering, the Department of Mechanical and Industrial Engineering also changed its name, when in 1989 it acquired the nuclear engineering program from the Department of Chemical Engineering. The Department of Mechanical, Industrial, and Nuclear Engineering (MINE) was granted its matching funds for an Ohio Eminent Scholar Chair in 1987, when Mrs. Ruth Alter donated $1 million in memory of her late husband, L.W. Scott Alter, who was president of the American Tool Company. August 1st of that year Atmaran H. Soni arrived at the college from Oklahoma State University to occupy the L.W. Scott and Ruth Alter Ohio Eminent Scholar Chair in Advanced Manufacturing, and to establish the Center for Advanced Manufacturing Systems.

Masterminding change in this department was Lee Harrisberger, who came out of retirement from the University of Alabama in 1989 to head MINE. Harrisberger was a pioneer in engineering education, recognized nationally for his concept of modules—mini courses in communication, leadership, interpersonal relations, group dynamics, and ethics. At UC he inaugurated on a grand scale his idea for the Design Clinic, a unique program for seniors to work in teams contracted to solve actual design problems in industry. In 1992 Harrisberger retired again, and was succeeded by Roy Eckart, a professor of nuclear engineering in the college since 1969.

By 1990 the faculty of the Department of Chemical Engineering had grown by five in as many years, under Sun-Tak Hwang, head of the department since 1982. One of the new faculty, Sotiris Pratsinis, was the first professor in the history of the college to win the National Science Foundation's prestigious Presidential Young Investigator Award in 1989. In the Membrane Technology Center, which was initially funded by Standard Oil of Ohio, Hwang formed the first of the college's industrial partnerships with an international cadre of companies from the United States, France, Japan, and Korea. It was the strength of this program that won state funding for the department for an Ohio Eminent Scholar Chair in Membrane Technology.

The growth of the college placed tremendous demands on the engineering library to become a state-of-the-art facility capable of providing the necessary resources for graduate education and research. The library's collection of journals more than doubled in six years, from four hundred titles in 1986 to nine hundred in 1992. The cataloging system evolved accordingly, from cards to electronic CD-ROM stations. Gifts from friends of the college, especially the Valentine, Bott, and Krug families, have provided for the library's collections. All of these improvements took place under Dorothy Byers, the college's first professional librarian, hired in 1972.

I was in band, waiting to get on the field for the half-time show, when this total monster of a human ran over me and my poor horn. My mouthpiece broke off the bottom of my tooth. Needless to say, I did not play in the halftime show, but I marched the whole thing, swallowing blood the entire time!"

–Lisa Bellin, M.E. '93

PROMINENT ARCHITECTURE *for* EMINENT RESEARCH

In the 1991-92 academic year the 137 full-time faculty members of the College of Engineering taught 2,000 undergraduate and 750 full-time graduate students. While the number of undergraduates has remained relatively stable over the past ten years, the full-time faculty roster has grown by 45 percent, and full-time doctoral students by 115% since 1985. Research expenditures have grown even more dramatically, by 250% since 1985, totaling $21 million in 1991-92.

To encourage the college's growth in research, the Ohio Board of Regents allocated $32 million in 1990 and 1992 for the design and construction of a new building for research and graduate education in sciences and engineering. As one of the first university research facilities constructed through state funds, the building represents the decision of the state and the Board of Regents that such facilities are important to Ohio's economic development.

In October 1990 the university selected KZF Incorporated in Cincinnati as the architect/engineer, and the Detroit-based firm of Smith, Hinchman & Grylls as laboratory design consultants. The university, the college, and KZF jointly selected Michael Graves, B.S.Arch. '58, D.F.A. '82, as the building's signature architect.

Dean Papadakis, right, and Michael Graves, left, discuss the site selection for the college's new building.

Graves, an internationally renowned architect, has been the Schirmer Professor of Architecture at Princeton University since 1962. Nationally, he has designed Riverbend Music Center in Cincinnati, the Humana Building in Louisville, the San Juan Capistrano Library in California, the Clos Pegase Winery in the Napa Valley, Disneyworld's Swan and Dolphin Hotels, and the Museum of Art and Archeology at Emory University. For the College of Engineering he has designed a terra cotta and ochre brick building with cast stone detailing. The vaulted roof, dormers, and stacks will be sheathed in copper. At the second and third levels the seven-story building will connect to the east end of Rhodes Hall. Its front entrance will face the corner of University Avenue and Woodside Place. It will provide 175,000 gross square feet of additional space for laboratories, faculty and graduate student offices, and conference rooms. Construction of the Graves building marks the beginning of a new phase of the college's history, yet at the same time it ties uniquely to the past. In 1894 another architect of Graves's stature, Samuel Hannaford, created the original McMicken Hall, the College of Engineering's first home.

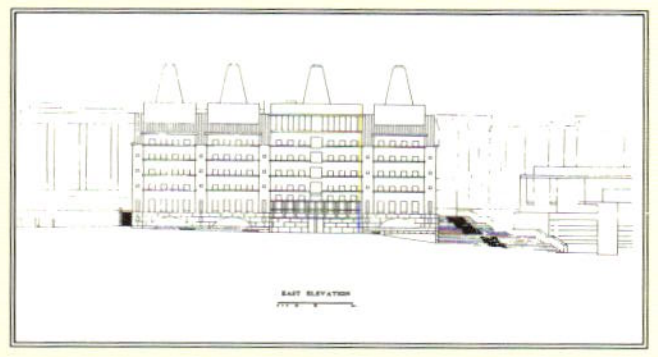

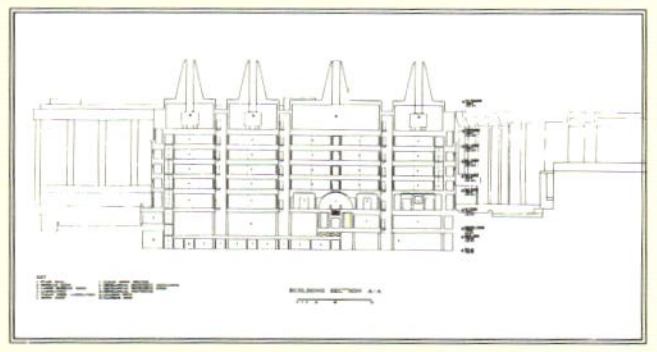

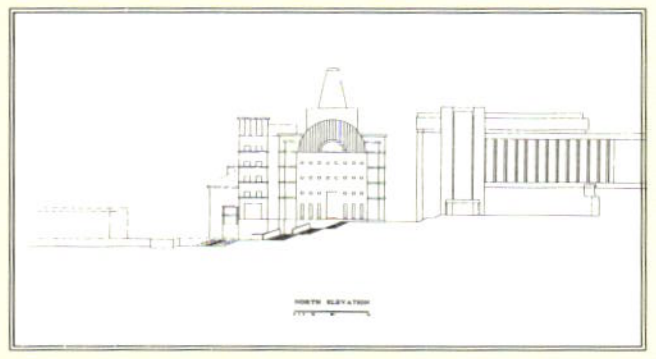

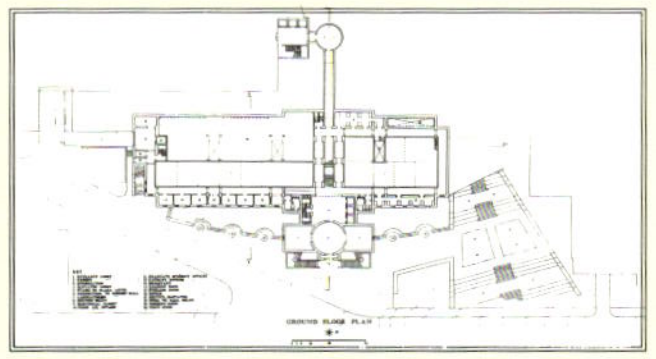

"I really enjoy the family-like atmosphere at the College. There are lots of laughs mixed in with the hard work." –Stuart Hooper, M.S. '94

Creating the physical facilities to accommodate the growth throughout the departments had been a constant challenge. The college acquired significant additional space in 1991 when Ronald Yocum, president of the Quantum Chemical Corporation USI Division, and a member of the College Advisory Council, orchestrated a plan for Quantum to lease its former research facility to the college for one dollar a year. Located seven miles north of campus, the facility provided sixty-five thousand square feet of space, including a highbay area and state-of-the-art wet laboratories, for several of the college's research centers and graduate programs.

On campus, preparations for a new science and engineering building, immediately east of Rhodes Hall, began in 1987 with extensive site studies in conjunction with the campus master plan. When the Ohio Board of Regents earmarked $25 million from its biennial capital appropriations for the building in 1990, the college was able to contract KZF Incorporated, of Cincinnati, and the Detroit firm of Smith, Hinchman & Grylls to design the structure. Michael Graves, B.S.Arch. '58, D.F.A. '82, was chosen by the college to be the signature architect. The Ohio Board of Regents appropriated another $7 million for the building in 1992, and the college launched its own capital fund drive to raise an additional $15 million. James A.D Geier, retired chairman of Cincinnati Milacron, continued his family's three-generation tradition of supporting the University of Cincinnati when he volunteered in September 1992 to recruit and chair the campaign committee.

The following November, Monarch Construction Company broke ground for the building, scheduled to be complete in 1994. At that time, the College of Engineering will house the past, present, and future in a unique way. Baldwin Hall is the oldest existing classroom building on the University of Cincinnati campus, and the Graves building will be, for some time, the newest. Inside those halls the College of Engineering stands ready to embark on its second century, showing no signs of slowing its drive for quality and innovation in education.

Top: After acceptance to the College of Engineering's baccalaureate Class of 1995, Andrea Flores, left, Landon Aldrige, middle, and Rhonda Taylor, right, attended the 1990 Bridge Program. Middle: As an undergraduate, Bina Soni, B.S.M.E. '92, took advantage of the college's state-of-the-art computing facilities. The IBM Workstation Laboratory is just one of several laboratories open to students twenty-four hours a day. Bottom: Stephen Clarson, a materials science professor, monitors a stress-strain apparatus which measures mechanical properties of elastomeric materials.

Epilogue

At the turn of the century, when three cities declined to try Herman Schneider's idea in their colleges and companies, no one could have predicted that cooperative education would eventually span the globe. In the United States alone, over nine hundred colleges and universities have adopted the Cincinnati Plan. Two hundred seventy-two schools in seventy-six other nations are now cooperative institutions.

In 1906 twenty-seven students in only three programs—electrical, mechanical, and chemical engineering—comprised the first class of co-ops. Today cooperative education belongs to almost every educational discipline. The fact that the number of business students has come to exceed the number of engineering students in cooperative programs in this country, and the fact that a Canadian school, the University of Waterloo in Ontario, offers the largest co-op program in the world, only affirms co-op's unlimited potential for innovation and adaptability.

In 1993 the University of Cincinnati is the world's third largest cooperative institution, after Waterloo and Northeastern University. As America's largest mandatory co-op program for engineers among public universities, UC continues to lead in other areas. The College of Design, Architecture, Art, and Planning offers the nation's largest co-op program in architecture, which is only one of an existing few. The university still influences cooperative programs nationally through the Center for Employer Training in Cooperative Education, a mobile program for employers of cooperative students. Since its groundbreaking inception in 1974 by Wanda Mosbacker, the associate dean for professional practice, the program's staff has traveled throughout the nation to help companies start co-op programs or to improve existing ones.

From the beginning, when they earned ten cents an hour, students have much appreciated the financial advantage of co-oping. In the late eighties and early nineties the University of Cincinnati alone has sent over twenty-one hundred students into the work force. Through co-op these students collectively gross over twenty million dollars to fund their education. Nationally the figure rises to billions of dollars—a sum which approaches what university financial aid offices, strained as they are, would otherwise have had to produce.

But more important than earning is learning. In many ways cooperative education is the ultimate scholarship. Students in co-op programs learn theories in the classroom and immediately apply them in practice. They acquire early sophistication with technology and insight into their chosen careers before committing to them. These benefits, which have prevailed from day one, provide the foundation for the outstanding educational opportunities offered at the University of Cincinnati College of Engineering.

APPENDIX CONTENTS PAGE

The Deans of the College of Engineering

Harry Thomas Cory
1900-1904

Harry Thomas Cory 1900-1904

Born in Lafayette, Indiana, on May 27, 1870. Bachelor's degree in mechanical engineering from Purdue University in 1887, and another, in civil engineering, from Purdue in 1889. Master's degrees in civil engineering and mechanical engineering, from Cornell, in 1893 and 1896, respectively. Professor of civil engineering at the University of Missouri from 1893 until coming to UC in 1900 to be the first dean of the College of Engineering. Engineered the closing of the break in the Colorado River in 1906, saving the Imperial Valley. Director of general foreign relief for the Red Cross during World War I. Chief engineer for irrigation in Seville, Spain, from 1925 to 1928. Died on March 22, 1955.

Thomas Evans
1904-1906

Thomas B. Evans 1904-1906

Born in Cincinnati, Ohio, on October 6, 1863. Bachelor's degree in chemistry from Columbia University in 1884. Doctorate in chemistry from the University of Erlangen, in Bavaria, in 1886. Chief chemist for the Proctor & Gamble Company from 1890 to 1892. Instructor of organic chemistry at the Massachusetts Institute of Technology from 1892 to 1895. Chief chemist for the American Cotton Oil Company from 1895 until coming to the University of Cincinnati in 1898. Began a program in technical chemistry in 1899, which formed the basis of the Department of Chemical Engineering. Became dean in 1904. Died on June 28, 1907, thus ending his career prematurely.

Herman Schneider
1906-1939

Herman Schneider 1906-1939

Born in Summit Hill, Pennsylvania, on September 12, 1872. Bachelor's degree in civil engineering from Lehigh University in 1892. Practiced architecture in Cumberland, Maryland, from 1894 to 1897, and then became a bridge engineer for the Oregon Short Line Railroad. Returned to be an instructor in civil engineering at Lehigh in 1899, where he began to formulate his plan of cooperative education. Came to the University of Cincinnati in 1903, and initiated his co-op plan the year he became dean, in 1906. Admitted the first women to the college in 1920. Served as president of UC from 1928 to 1932. Died in office on March 28, 1939.

Robert Clyde
Gowdy
1939-1946

Robert Clyde Gowdy 1939-1946

Born in Springfield, Ohio, on March 10, 1886. Earned all of his degrees, in physics, from UC: a bachelor's in 1906, a master's the following year, and a doctorate in 1909, at age twenty-three. Instructor of physics at Lehigh University for one year before coming to UC in 1912. First lieutenant for the Ordnance Department and a captain in the Chemical Warfare Service during World War I. Assumed the dean's duties during Herman Schneider's presidency. Steered the college through financial drought when the student population plummeted during World War II. Awarded an honorary doctor of science from UC in 1947, one year after he retired. Died March 27, 1950.

C. Albert Joerger
1947-1957

Charles Albert Joerger 1947-1957

Born in Mulhouse in the Alsace region of France on October 6, 1886. Earned the degree of mechanical engineer from the Stevens Institute of Technology in 1910. Instructor at Harvard University for one year before becoming an instructor in mechanical engineering at UC in 1913, and eventually department head. Became dean in 1946, the first year that the college resumed its original title when it separated from the Colleges of Commerce and Applied Arts. Managed the overwhelming enrollment of World War II veterans attending the College of Engineering under the GI Bill. Admitted the first minority students in 1950. Died on April 6, 1971.

Howard Kenneth
Justice
1957-1962

Howard Kenneth Justice 1957-1962

Born in Williamstown, Kentucky, on January 6, 1896. Member of UC's civil engineering Class of 1919. Lettered in football, basketball, and track. Doctorate in mathematics from UC in 1933. Began teaching at UC in 1921 as an instructor in mathematics. As dean, oversaw the initiation of the drive to build the college into a first-rate graduate institution. Returned to teaching in 1962 before retiring in 1966. Honored as a Distinguished Alumnus in 1973. With failing eyesight, he completed and published his mathematical proof of a spherical surface in the earth-moon gravitational field at age ninety-six, in the College Mathematics Journal of the Mathematical Association of America in 1992.

Cornelius Wandmacher
1962-1974

Cornelius Wandmacher 1962-1974

Born on September 1, 1911. Bachelor of civil engineering from the Polytechnic Institute of New York in 1933, and a master's degree two years later. Came to UC in 1951 from teaching civil engineering at his alma mater. Initiated Recognition Day in 1963 and the Distinguished Alumnus Award in 1969. Returned to the classroom from the dean's office in 1974, and continued to hold leadership positions in national organizations. President of the American Society of Engineering Education from 1974 to 1975. Member of the Board of Directors of the American National Metric Council from 1977 to 1982. Retired from teaching in 1981. Died on September 12, 1992.

Clyde Melvin Adams, Jr.
1974-1976

Clyde Melvin Adams, Jr. 1974-1976

Born on August 13, 1925, in Detroit, Michigan. Served in the US Naval Construction Battalions in the Pacific theater from 1943 to 1946. Earned a bachelor's in 1955, a master's in 1957, and a doctorate in 1963, all in metallurgy, and all from the Massachusetts Institute of Technology. Faculty member at MIT for sixteen years. Came to UC in 1974 from the University of Wisconsin-Madison. International consultant for industrial and insurance companies, research laboratories, and government agencies. Relinquished his position as dean in 1976 to pursue research and development, and to teach in the field of energy.

Robert Mitchell Delcamp
1976-1978

Robert Mitchell Delcamp 1976-1978

Born in Lexington, Kentucky, on April 18, 1919. Bachelor's degree in chemistry from Transylvania University in 1939. Both graduate degrees in organic chemistry from the University of Cincinnati: a master's degree in 1941 and a doctorate in 1954. Chemist and later research associate in the Tanners' Council Research Laboratory from 1954 to 1968. Taught organic chemistry in the college from 1945 to 1965. Assistant dean from 1962 to 1967. Continuously involved in student affairs, including the Engineering Tribunal and The Cooperative Engineer. Acting dean of the College of Engineering from 1976 to 1978. Elected fellow of the Ohio Academy of Science in 1978. President of the Engineering Society of Cincinnati from 1978 to 1979.

Kirk C. Valanis 1978-1983

Born on March 6, 1930, in Lefkara, Cyprus. Graduated in aeronautical engineering from Imperial College, London, with a bachelor's in 1955 and a master's in 1957. Doctorate in aeronautical engineering from Purdue University in 1963. Professor of aeronautical and engineering sciences, engineering mechanics, and chemical and materials engineering at Purdue, Iowa State University, and the Stevens Institute of Technology. Chaired the Division of Materials Engineering at the University of Iowa before coming to the University of Cincinnati in 1976. Founding member of the Society of Engineering Science and a consultant to the army and to industry.

Kirk C. Valanis
1978-1983

Louis McNeal Laushey 1983-1986

Born May 13, 1917, in Columbia, Pennsylvania. Bachelor of science in civil engineering from Pennsylvania State College in 1942. Master of science in civil engineering and a doctor of science in engineering in 1947 and 1951, respectively, from the Carnegie Institute of Technology. Professor and head of the Department of Civil Engineering at Norwich University from 1954 to 1958. Board of Directors of the American Society of Civil Engineers from 1981 to 1984, and vice-president from 1984 to 1986. William Thoms Chair of Civil Engieering at UC from 1958 to 1978. Executive officer and then interim dean of the College of Engineering from 1983 to 1986.

Louis McNeal Laushey
1983-1986

Constantine N. Papadakis 1986-Present

Born on February 2, 1946, in Athens, Greece. Diploma in civil engineering from the National Technical University in Athens in 1969. Master's degree in civil engineering from the University of Cincinnati in 1970, and a doctorate in hydraulics from the University of Michigan in 1973. Taught civil engineering at George Washington University, the University of Michigan, and Michigan State University. Assistant chief engineer at Bechtel Associates in Ann Arbor from 1978 to 1981. Vice-president at STS Consultants in Ann Arbor from 1981 to 1984. Vice-president at Tetra Tech, Inc., in Pasadena in 1984. Chaired the University of Colorado Department of Civil Engineering from 1984 to 1986.

Constantine N. Papadakis
1986-Present

The Faculty of the College of Engineering

Listed by date of first appearance in a college bulletin, which is usually one or two years after appointment level, ranging from student assistant to dean.

1875
Henry Turner Eddy, Dean,
Astronomy, Mathematics, Civil
Edward W. Hyde,
Mathematics, Civil

1880
Ward Baldwin, Mathematics, Civil

1881
Nelson W. Perry, Metallurgy

1883
John B. Porter, Metallurgy

1893
Louis E. Bogen, Physics, Civil

1894
Joseph B. Strauss, Civil

1895
Philo A. Orton, Civil

1896
William C. Bauer, Civil

1897
Willard C. Walton, Civil

1898
George C. Thomas, Civil

1900 - 1901
Edward M. Brown, English
Harry T. Cory, Dean
Thomas Evans,
Technical Chemistry
Harris Hancock, Mathematics
Jermain G. Porter, Astronomy
Louis T. More, Mathematics
Stephen E. Slocum, Applied
Mathematics, Civil
Frederick H. Safford,
Mathematics, Physics
Louis E. Bogen, Physics
Frank I. Sheperd, Chemistry
Florence C. Lawler, Mathematics
John T. Murray, English
Azariah Lincoln, Chemistry
George L. Hamilton,
Romance Languages
Horace D. Chipman, Shopwork

1902
Frank O. Dufour, Civil
Horace T. Eddy, Electrical
Harry S. Fry, Chemistry
Robert R. Harkins,
Machine Design and Drawing
James E. Ives, Physics
Carl A. Krause, Romance Languages
Christian W. Marx, Mechnical
Robert Promberger,
Forging, Machine Shop
Sylvanus C. Shipley, Woodwork
John F. Snell, Chemistry

1903
Clement Bell, Woodwork, Forging
William B. Cartmel, Physics
Edwin G. Greenman,
Machine Design and Drawing
Phillip H. Harry, Romance Languages
Leigh Rodgers, Wood Shop
Herman Schneider, Civil
John H. Traeger, Civil

1904
William Cartmel, Physics
Earl F. Farnau,
Chemistry, Mathematics
Arthur Garlick, Forging
Robert C. Gowdy, Physics
Frederick C. Hicks, Economics, Civics
Alexander Jenkins, Mechanical
Alexander C. Lanier, Electrical
Marco Liberma, Romance Languages
Claude M. Lotspeich, German
Elizabeth Merrill, English
George Miller, English
Henry B. Phillips, Mathematics
Max Poll, Germanic Languages
Henry Schilling, Woodwork
Elsie Schrader, German
Delisle Stewart, Astronomy
George W. Umphrey,
Romance Languages

1906
Samuel Allen, Physics
Fred Ayer, Civil
Edward Barrier, Analytical Chemistry
Harris M. Benedict, Biology
John Davis, Physical Education
John T. Faig, Mechanical
George Johnston, Chemistry
Arthur Kinsella, Greek, English
Charles Underwood,
Romance Languages
Ralph Warrin, Civil

1907
Louis Brand, Mathematics
Louis Brockhaus, Civil
Robert Brown, Mechanical
Earl F. Farnau, Chemistry
Nevin M. Fenneman,
Geology, Geography
Michael Guyer, Zoology
Edward Hatterschide, Chemistry
Ralph Inott, Physical Education
Lauder W. Jones, Chemistry
Joseph H. Kindle, Mathematics
Ralph O. Oesper, Chemistry
Clyde W. Park, English
John Porter, Metallurgy
Bernard Scharr, Chemistry
Peter Sivas, Physical Education
Sherman Turrill, Civil
Winfred Webber, Mathematics
Clarence R. Wylie, Electrical

1908 - 1909
Robert Brooks, Political Science
Robert Brown, Mechanical
Taylor S. Carter, Physics
Howard A. Dorsey, Mechanical
Clarence Eckert, Chemistry
Fayette Elwell, Economics
Clifford Fels, Chemistry
Charles Hecker, Chemistry
Max Isaac, Physical Education
Amor Kistner, Civil
George Krapp, English
William Lauder, Chemistry
Charles N. Moore, Mathematics
Edward Rieman, Physics
Joseph E. Root, Civil
Samuel N. Taylor, Electrical
Winfred Webber, Mathematics
Raymond Witte, Physical Education

1910
Samuel Allen, Physics
Ernest Carman, Geology
Frank Chandler, English
James Dilworth, English History
Harry Fry, Chemistry
Carl Knoebel, Mechanical
Henry Kranz, Civil
Edward Lorenz, Physics
George Miller, English
Louis T. More, Physics
Curtis C. Myers, Coordination
Van Schlicten, Chemistry
James Trimble, Mathematics
Fred Upson, Chemistry
Benjamin C. Van Wye, Public
Speaking, English
Merrick Whitcomb, History

1913
James Aston, Metallurgy
G. M. Braunne, Civil
Alfred Brodbeck, Physical Education
Charles W. Brown, Labratory Arts
Harold Collins, Mechanical
Isaac Cox, History
Edwin W. Esslinger,
Analytical Chemistry
Charles B. Hoffmann, Electrical
Charles Albert Joerger, Mechanical
Joseph Kindle, Mathematics
Selden Lowrie, Political Science
Martin Ludwich, French, German
James Magee, Economics
George Moore, Civil
Clarence Nash, Physical Chemistry
Max Robinson, Coordination
Stephen Slocum,
Applied Mathematics
Edward Smith, Mathematics
Clarence Stevens, English
Alexander M. Wilson, Electrical
Russell Witte, Civil

1914
Otto Birk, English
Hubertis Cummings, English
Albert Davison, Physical Chemistry
Clarence Fernberg, English
John Long, Civil
Cannon Sneed, Chemistry
Ernest Thum, Metallurgy

1915

Walter H. Bucher, Geology
Albert Davidson, Physical Chemistry
R. G. Knickerbocker, Metallurgy
John Long, Civil
Phillip Ogden, Romance Languages
Walter Plock, English

1916

Ion Cortright, Director of Athletics
Myers, Curtis, Industrial
Amos Herold, English
Percy K. Holmes, Physical Education
Harold Janda, Civil
James Orr, Physical Education
Edward K. Ruth, Civil
Edwin Stenger, Metallurgy
Jesse Williams,
Hygiene, Physical Education

1917

E. A. Eberhardt, French, German
Victor Mavity, Electrical
Whitelaw Morrison,
Physical Education
Warner Tabb, Electrical
Charles Wood, Metallurgy

1918

Charles Carroll, Civil
Boyd Chambers, Athletics
Stephen Felton, Metallurgy
W. F. Lewis, ROTC
Fred Molby, Physics
A. H. Morse, Production Engineering
Ralph Oesper, Analytical Chemistry
William C. Osterbrock, Electrical
Clifford Rolle, Chemistry
Frederick Schroder, Civil
Bartholomew Shine, Mechanical
Orland Sweeney,
Industrial Chemistry
Charles Wood, Metallurgy
George Zugelter, Mechanical

1919

Lowell Alexander, Physics
Edgar Bengert, English
Elmer Coleman, Electrical
W. H. Creighton, Laboratory Arts
Vincent H. Drufner, Coordination
Geoffrey Gray, Chemistry
Ralph L. Langenheim, Civil
Reginald C. McGrane, History
H. Calvert Messinger, Coordination
Fred Molby, Physics
Thomas Mullin, Civil
Fay Norton, Coordination
Joseph O'Conner, Chemistry
Merwyn Rodi, Electrical
Meyer Salkover, Mathematics
William Willard, Industrial
Henry Wolsdorf, Mechanical

1920

Gerald Barnes, Physical Education
Edward Bengert, English
Wallace W. Boone, Metallurgy
George Burns, Coordination
Earl Case, Geography
Alfred Daum, Labratory Arts
Howell Gano, Industrial Management
Edgar Gilman, Civil
Walter Haehnle, English
Arthur Hathaway, Civil
Eric Lane, Chemical
Edward Lorenz, Physics
C. H. Lund, Coordination
Newton Noble, Electrical
Alexander Paterson, Civil
Carlton Smith, Chemical
Marjorie Stewart, Coordination
Erwin Tschudi, Mathematics
Edgar VanBuskirk, Hygiene
Keiskell Whaling, Economics
Nathaniel Whitney, Finance
Leroy Wilson, Mechanical

1921

Rupert A. Anderegg, Civil
Daniel Cook, Sketching
E. W. Detraz, Metallurgy
Harvey Drach, English
Clarence Fackler, Commerce
Hal Fikret, Mechanical
Robert E. Hundley, Mathematics
Howard K. Justice, Mathematics
Arthur Knebel, Mechanical
George McLaughlin, Leather Research
Harry L. Miller, Mathematics
Walter Muhlbach, Commerce
C. L. Pfeiffer, Electrical
Leo Raub, Physics
Alfred Ribbink, Commerce
George Rockwell, Leather Research
Edwin Theis, Leather Research
Reuben S.Tour, Chemical
William S. Wabnitz, English
Karl Wecker, Music

1922

Griffith T. Addison, Coordination
Walter A. Baude, Coordination
Herbert A. Dangel, Mathematics
Clarence Johns, History
William Johnston, Geology
Frederick Jordan, Marketing
Howard B. Luther, Civil
Stanley Matthews, Architecture
Harold Sibert, Aeronautical
Wayne Steele, English
Charles W. Theis, Geology

1923

Cecil Baker, Architecture
Alexander Blair, History
Wilfred Bunker, English
Wilbur Calhoun, Finance
Charles Dwight, Physics
E. G. Ebertz, Leather Research
George M. Enos, Metallurgy
Platt Evens, Laboratory Arts
Norwood C. Geis, Commerce
Alfred Geisler, Law
James Harland, Classics
Ernest B. Haswell, Architecture
Clarence Lubin, Mathematics
Lewis F. Palmer, Mathematics
Albert Savoy, Mathematics
John W. Surbaugh, Mathematics
Robert Taylor, Commerce

1924

Saul Arenson, Chemical
Donald Beddoe, English
Charles Behre, Coordination
Lawerence Culver, Electrical
Armand Knoblaugh, Sound Research
Carroll Leonard, Mechanical
Roy McDuffie, Metallurgy
Howard Martin, Geography
Muriel Metz, Mathematics
Frederick O'Flaherty, Leather Research
Robert F. Reed, Lithographic Research
Reed F. Stockdale, Coordination
Henry Twitchell, English
Leonard Watkins, Economics
Dare Wells, Physics

1926

Francis H. Bird, Commerce
Manson Bradley, Chemical
J. W. Bunting, Mechanical
Wilbur Calhoun, Finance
Samuel Fife, Mathematics
John T. Geary, Military Science
and Tactics
Harold J. Kersten, Mathematics
John Loofbourow, Physics
Marcus Mullings, Mathematics
John Oddy, History
William Russ, History
Wilfred Schubert, Coordination
R. R. Sharrock, Marketing
George Sperti, Research
Ralph A. Van Wye, Coordination

1927

Edward Barta, Chemical
Daniel R. Bergsmark, Geography
Tabitha Cobb, Biochemistry
Herbert Feis, Economics
Clarence Gardner, Political Science
Hoke S. Greene, Chemical
Arthur Holmes, Commerce
Ralph Langenheim, Coordination
George McLaughlin, Leather Research
Helen Moore, Chemistry
Harold Patton, Economics
Melrose Pitman, Women's Work, Art
James A. Quinn, Sociology
Frederick W. Rogers, Mathematics
Guy A. Tawney, Ethics
I. David Wollermann, English

1928

Ward E. Bullock, Mechanical
Earle Eubank, Sociology

Raymond Renn, Civil
Leslie J. Schwallie, Coordination

1929
George Babcock, Physical Training
Isay Balinkin, Physics
William Hewett, Economics
Bradley Jones, Aeronautical
Reginald McGrane, History
Ralph Miller, Coordination
Adolph Sandberg, Geology
Reuel L. Smith, Mechanical

1930
Earl Aylstock, Commerce
Julian Bechtold, Modeling
Olive Betts, Interior Decoration
Myrl Bottomley,
  Landscape Architecture
Daniel Cook, Sketching
Carlotta Corpron, Design
Russell Dunholter, Mechanical
John Earhart, Theory of Color
Annette Edens, Art Structure
Josephine Fulks, Applied Arts
Reginald Grooms, Freehand Drawing
Henry Kenney,
  Landscape Architecture
Alan Laing, Architectural History
Alice Little, Applied Arts
Henry Marley, Ceramics
Leslie Martin, Architectural Design
Harold Nash, Ceramics
Oliver Nickoloff, Physical Training
Ernest Pickering, Architectural Design
Jessie Roberts, Art Structure
George Roth, Architecture
Josephine Swift, Art Structure

1931
Arthur Kelsey, Architectural Design
Harold Vinacke, Political Science

1932
Muriel Ater, Mathematics
James Ervin, Mathematics
Carl Gieringer, Electrical
Harry Henig, Economics
Robert Hynes, Coordination
Shelden G. Lowrie, Political Science

John Rich, Geology
Shuey Wolfe,
  Military Science and Tactics

1933
Roy Blough, Economics
Ward Bullock, Mechanical
Allan Chace, Laboratory Arts
John Keto, Electrical
Hope Warner, Art

1934
Ralph C. Bursiek, Commerce
Albert C. Herweh, Electrical
Edward F. White, Mathematics

1935
Robert Harris, Political Science
Henry Hodges, Political Science
John Jackson, Mechanical
Edwin T. Miller, Mathematics
Mortimer Powell, Aeronautical
John Weaver, Chemical

1936
John L. Baker, Mathematics
Bruce B. Baldwin, Mechanical
Hollis Barber, Political Science
George Barbour, Geology
George F. Deasy, Geography
Harold J. Garber, Chemical
Herbert Owens, Mathematics
Francis Rich, Mathematics
William Vogel, History
Thomas Zandstra, Mathematics

1937
Arthur Holmes, Accounting
Adolph Sandberg, Geology

1938
Kenneth Hittinger, Commerce
Roy Marz, English

1939
William Thomas, Accounting

1940
Alexander Bereskin, Electrical
Wayne Felts, Geology
John F. Kahles, Metallurgy
Kaiser Kunz, Mathematics

William Licht, Chemical
Hans Liepmann, Aeronautical
William E. Restymeyer, Electrical

1941
Jacob Cholak, Research
Russell Chrysler, Marketing
Charles Higgins,
  Military Science and Tactics
Gabriel Horvay, Electrical
Willy Lange, Applied Science
Hans Liepmann, Aeronautical
Urban Linehan, Geography
Carl Ludeke, Mathematics
George Rieveschl, Jr., Chemistry
Landon G. Rockwell, Political Science
Robert Russell, Mineralogy
Carl Seddon, Electrical
Clarence Simpson, English
Walter Soller, Applied Science
Frank Tiller, Chemical
Philip Wallace, Mathematics

1942
Raymond L. Bisplinghoff, Aeronautical
James Black, Applied Science
William W. Hagerty, Mechanical
Samuel H. Manian, Chemical

1943
Joshua Barlaz, Mathematics
Elmer Bodman, Mechanical
Carl W. Condit, Mathematics
Dwight E. Crummer, Mechanical
Jules J. Fern, English
Robert Hoefer, Accounting
H. David Lipsich, Mathematics
William Reynolds, Applied Science
Ele E. Stansbury, Metallurgy
Jean Winston, Mathematics

1944
Jay H. Clow, English
Harold Hancock, Applied Science
Donald Montgomery, Electrical
Wanda B. Mosbacker, Coordination
Robert Ramey, Applied Science

1945
Don Gebhart, Mechanical
E. S. Furnish, Military Science and Tactics

1946
Samuel Allen, Experimental Physics
Robert Delcamp, Organic Chemistry
Louis F. Doty, Aeronautical
William H. Hoblitzell, Coordination
James O'Connell, English
Carl Schneider, Applied Science
John Wicklatz, Applied Science
James Vaughn, Psychology

1947
Philip Berghausen, Applied Science
Martin Berman, Mathematics
William Blaylock, Physics
Ina Bryner, English
Walter Buchman, Coordination
Albert J. Cornish, Physics
Harold F. Feldstein, Mathematics
Allan B. Folger, Mechanical
Robert Fopma, Mathematics
Alex Green, Physics
Richard Hanford, Physics
William Harlow, Mathematics
William Hawley, English
Robert T. Howe, Mathematics
Grover F. Hubing, Physics
Richard B. Humphreys,
  Engineering Drawing
Rowland Kleine, Electrical
Reuben Koch, Mathematics
Theodora Kohler, English
Lloyd W. Krogh, Mathematics
John Krueger, Applied Science
Robert Lollar, Tanning Research
Lewis Miller, Chemical
Lawrence Morgan, Chemical
John Nader, Physics
George Nardin, Accounting
Dennis Nead, Mathematics
Herman Newman, English
H. S. Osgood, Inorganic Chemistry
F. William Pape, Physics
Edward T. Price, Geography
Gordon Rittenhouse, Geology
Jacob H. Sarver, Engineering Drawing
Robert Schulte, Electrical
Charles Schultz, Coordination
William M. Spurgeon, Applied Science
Flora Stoehr, English

Louis Steffens, Engineering Drawing
Gerhard Vater, Mathematics
Hugh Wales, Marketing
William Walker, Mechanical
Albert Wellman, Mathematics
Sherrell Wilkes, English
Lan J. Wong, Aeronautical
Joseph A. Zumberg,
Engineering Drawing

1948

August Biederman,
Engineering Drawing
Charles Borders, Electrical
H. C. Brinkman, Mathematics
Charles Buch, Chemical
John Dedrick, Metallurgical
Russell J. Dickason, Electrical
George Engberg, History
Francis E. Espelage, Laboratory Arts
Ralph W. Fowler, Electrical
Clair Hubert, Psychology
Charles E. Huckaba, Chemical
Clair Hulley, Mechanical
Vernon Jolley, Accounting
Gerard Kraus, Applied Science
Charles Koch, Economics
James King, English
Leigh Matthews, Electrical
William H. Middendorf, Electrical
Amy Pathe, Business Correspondence
Cecil Phillips, Engineering Drawing
Eric Wang, Mathematics

1949

Arthur Bills, Psychology
Murvel J. Bratt, Mathematics
Donald L. Crews, Metallurgical
John Dewey, Chemical
Richard H. Engelmann, Electrical
Rudolf Feige, Mathematics
Charles Frank, Applied Science
Joseph C. Green, English
Chester Jackson, Mechanical
Arthur C. Nolte, Aeronautical
J. Edward Oetjen, Mechanical
Garland G. Parker, History
Robert Price, Chemistry
George Shelby, Economics
Fred E. Westermann, Metallurgical

1950

William A. Hendricks, Mechanical
Harold Hoelscher, Chemical
Arthur Knebel, Engineering Drawing
Wayne Overmyer, Accounting
E. C. Pattee, Chemical
James Quinn, Sociology
Richard Stevens, Mathematics

1951

L. M. Alexander, Physics
Theodore Anderson, Finance
Clark Aumend, Marketing
Harrison Dwight, Physics
Carl Evert, Electrical
Maurice G. Frey, Geology
Gerald M. Friedman, Geology
Milan Karas, Marketing

1952

Robert P. Borofka, Electrical
Donald Crews, Metallurgical
Paul M. Gray, Electrical
Wilton F. Melhorn, Metallurgical
George D. Moon, Inorganic Chemistry
William Studt, Coordination
Cornelius Wandmacher, Civil

1953

George Cunningham, Applied Science
John Dickman, Laboratory Arts
J. R. Gump, Chemistry
Herbert Koch, Business Administration
Robert Lemlich, Chemical
Roy H. Murray, Aeronautical
Robert H. Rakel, Electrical
Harvey C. Sunderman, Geology

1954

Nathan Gilbert, Chemical
Dale Jones, Chemical
Ralph E. Peters, Coordination
Bruce Raymond, Aeronautical
Daniel J. Schleef, Mechanical
William Tholke, Metallurgical
Nelson T. Williams, Chemistry

1955

Robert E. Dillon, Marketing
Charles Dittrich, Air Science
Richard Durrell, Geology

Odin Elnan, Aeronautical
Chad F. Gottschlich, Chemical
Charles Griffin, Accounting
Ralph M. Knapp, Coordination
Irvin L. Lynn, Mathematics
Robert Martin,
Military Science and Tactics
David Mazkewitsch,
Engineering Drawing
James Sheehan, Marketing

1956

Albert G. Bruck, Mechanical
William Jenks, Geology
Dale Kiefer, Accounting
Gaylord Merriman, Mathematics
Gordon Skinner, Labor Relations
William Vogel, History
William Williamson, Air Science

1957

Andrew R. Barkocy, Civil
James Conkin, Geology
Marvin L. English, Mechanical
John D. Eye, Environmental
Donald E. Hagedorn,
Engineering Graphics
Michael Hoch, Metallurgical
John T. McCarthy, Physics
Charles Oligee, Mechanical
Carl H. Osterbrock, Electrical
William Pape, Physics
Arnold Schreier, History

1958

Ted Cook, Mathematics
Karl H. Matthies, Mathematics
Harry C. Saxe, Civil

1959

Hans Ernst, Research
Thomas A. Gallagher, Mathematics
Clifton Hazard, English
C. G. Hubbart,
Military Science and Tactics
Robert J. Kroll, Aeronautical
Louis M. Laushey, Civil
Gene D. Lewis, History
Charles Pinzka, Mathematics
Arthur C. Plate, Engineering Graphics
Arnold Polak, Mathematics

Widen Tabakoff, Aeronautical
Richard Williams, Engineering Graphics

1960

William S. Clark, English
Harry Ferguson, Mathematics
George Gregg, Air Science
Paul Harrington, Aeronautical
Arnold Polack, Mathematics
Ronald Schmidt, Geology

1961

Andrew Bodocsi, Civil
Max L. Brown, Mechanical
Herbert F. Curry, History
Williams Foster, Engineering Graphics
Theodore W. Gilbert,
Analytical Chemistry
Richard G. Griskey, Chemical
Hilmar Krueger, History
Larry R. Last, Electrical
Joseph Loch, Aeronautical
Wesley Love, Mathematics
Kenneth Meyer, Mathematics
Leland Miles, English
John D. Mulholland, Aeronautical
Boyd C. Ringo, Civil
Charles Saltzer, Applied Mathematics

1962

Howard T. Anderle, Mathematics
Victor Dial, Speech
Ralph L. Fairchild, Mathematics
Arthur Focke, Metallurgical
William F. Grayburn, English
Thomas Hambrick, English
Daniel Havens, English
Truman F. Keefer, English
Ting Yi Li, Aerospace
Lloyd Moote, History
Ivan E. Morse, Mechanical
Russell F. Osborn, Aerospace
Michael Porte, English
Benjamin A. Sokoloff, English
Peter Topping, Later Greek Studies
Yeong-Ging Tsuei, Mathematics
William S. Wagner, Electrical
William R. Wright, Physics

1963

Emil Dansker, English
Violet Diller, Biophysics
Paul Evans, Coordination
William Foster,
  Engineering Graphics
Ernst Franke, Biophysics
Myron Funk, Military Science
William E. Gates, Sanitary
Daniel Hershey, Chemical
Hans Hoffmann, Geology
Ronald L. Huston, Mechanical
Amolak C. Jain, Aerospace
Dean Johnson, English
Dale Jones, Physics
Frank Kafker, History
Jason R. Lemon, Mechanical
Khalil H. Mancy, Sanitary
Bertram Mond, Mathematics
Harvo Oguro, Aeropspace
Maido Saarlas, Aerospace
Solomon Schwebel, Physics
James T. Scroggin, Mechanical
James E. Sherlock,
  Engineering Graphics
Dean S. Shupe, Mechanical
Kuo Wang, Electrical
Thomas C. Ware, English
Helmut Weik, Physics
George Winter, English

1964

Jesse E. Bybee, Electrical
Stanley L. Cosgrove,
  Organic Chemistry
Charles H. Fitts, Mathematics
William A. Grissom, Mechanical
Thomas W. Gurbach,
  Engineering Graphics
Stephen D. Harden, Mathematics
Ralph Knapp, Coordination
George Lee, Aerospace
Harvey E. Maseck,
  Engineering Graphics
Frank Rizzo, Metallurgical
Pasquale V. Scarpino, Sanitary
Richard A. Scholten, Electrical
Francis S. Tse, Mechanical
Robert Weise, Coordination

1965

Ralph G. Dale, Aerospace
Thomas Gurbach,
  Engineering Graphics
Paul V. Guthrie, Metallurgical
J. Vaughn Hackworth, Metallurgical
Anis Kanjo, Mathematics
James M. Morand, Civil
Herbert C. Preul, Civil
Alfred W. Scheide, Electrical
Brent Stojkov, Mechanical
Donald W. Sundstrom, Chemical
Frank R. Tepe, Aerospace
Ronald Zerkle, Mechanical

1966

William J. Dorson, Chemical
Rodney P. Elliott, Metallurgical
David A. Fiebig,
  Engineering Graphics
Kenneth H. Flint, Coordination
Willliam Foster, Engineering Graphics
James H. Leonard, Nuclear
James F. McDonough, Civil
John Martin, Coordination
Philip Marvin,
  Business Administration
Gary Slater, Aerospace

1967

Carl F. Abegg, Chemical
William Bunce, Coordination
Ronald D. Early,
  Engineering Graphics
Leroy Eckart, Mathematics, Nuclear
Reinhold Erickson, Mechanical
Harold Franklin, Mechanics
Charles W. Gruber, Environmental
Albert L. Klosterman,
  Engineering Graphics
Jason Lemon, Mechanical
Billy O. Mullins, Electrical
Robert L. Myers, Mathematics
John N. Pattison, Environmental
Arnold Polack, Aerospace
Alan J. Rubin, Civil
James Sheng, Aerospace
James Sherlock, Mechanical
John F. Stephens, Mechanical
Bronson Sweeney,
  Engineering Graphics
I-Chih Wang, Mechanical

1968

Edward W. Allis, Mechanical
George E. Beine, Electrical
Gerald Berg, Civil
Frank E. Burris, Electrical
John P. Cook, Civil
John Coy, Mechanical
Sean P. Flanagan, Engineering Graphics
William Grissom, Mechanical
Ram Kavipurapu, Nuclear
Joseph N. Lockhart, Mathematics
Leonard Meirovitch, Aerospace
Roger E. Messick, Mathematics
Robert Myers, Mathematics
Joseph H. Nevin, Electrical
Robert H. Raible, Electrical
James Scroggin, Electrical
Alvin Shapiro, Nuclear
William Shapton, Mechanical
Bronson Sweeney,
  Engineering Graphics
James F. Thorpe, Mechanical
Gerald G. Ventre, Aerospace
William Wagner, Electrical

1969

Stephen D. Antolovich,
  Materials Science
David J. Barr, Civil
James F. Brown, Mechanical
Wilmer Clevenger, Aerospace
John Coy, Mechanical
H. Thurman Henderson, Electrical
Charles Keaney,
  Engineering Graphics
Riley N. Kinman, Civil
John J. McDonough,
  Engineering Graphics
Chris E. Passerello, Aerospace
Richard A. Pistler, Mechanical
Monte J. Pool, Materials Science
Robert H. Price, Materials Science
Casimir J. Rosa, Materials Science
Donald E. Sundgren,
  Engineering Graphics
Nicholas Vitale, Aerospace
Gerald H. Weidenhamer, Aerospace
Joel Weisman, Nuclear
William R. Wells, Aerospace

1970

T. Michael Basehart, Civil
Robert A. Calico, Aerospace
Richard P. Chartoff, Chemical
Theodore R. Comstock, Mechanical
Stanley Day, Electrical
Neal Des Ruisseaux, Mechanical
Kirti N. Ghia, Aerospace
Chester K. Greathouse, Mechanical
Neal P. Jefferies, Mechanical
James R. Luginbuhl, Mechanical
John Moteff, Materials Science
Robert H. Nilson, Mechanical
Donald C. Stouffer,
  Engineering Science
Donald Vieth, Metallurgical
William R. Zimmerman,
  Mechanical

1971

James N. Anno, Nuclear
John M. Christenson, Nuclear
John Dumbleton, Materials Science
Charles D. Fournier, Chemical
Chester Greathouse, Mechanical
Issam A. Minkarah, Civil
Alvin M. Strauss, Engineering Analysis

1972

Neil A. Armstrong, Aerospace
F. James Boerio, Materials Science
Marcelo Crespo da Silva, Aerospace
R. Thomas Davis, Aerospace
William Kauffman, Aerospace
William G. Wee, Electrical
Michael J. Werle, Aerospace

1973

Awatef Hamed, Aerospace
Lawrence H. Sobel, Aerospace

1974

Joseph T. Boyd, Electrical
Urmila Ghia, Aerospace
James Klemm, Engineering Analysis

1975

Clyde Melvin Adams,
  Engineering Education
David B. Greenberg, Chemical
Howard Siegel, Engineering Analysis
E. Sam Sovilla,
  Professional Development

1976

Randall J. Allemang, Mechanical
Georgés Becus, Engineering Science
Yuen Koh Kao, Chemical
Soon Jai Khang, Chemical
Eric Malstrom, Mechanical
Adnan H. Nayfeh, Aerospace
Janet Rickabaugh, Environmental
Ryong-Joon Roe, Materials Science
Richard Shell, Industrial
Howard Siegel, Engineering Analysis
Frederick J. Taylor, Electrical

1977

Ian A. Davidson, Electrical
Johannes Penninger, Chemical
Richard W. Young,
  Engineering Science

1978

Kenneth Meyer, Mathematics
Carl Panek, Engineering Science
Lazar Papernik, Engineering Science
David Richardson,
  Engineering Science
Kirk C. Valanis, Electrical

1979

William J. Atherton, Mechanical
Kenneth M. Emmerich, Nuclear
Louis W. Trent, Professional Practice

1980

Joel R. Fried, Chemical
Rakesh Govind, Chemical
Prem K. Khosla, Aerospace,
  Applied Mechanics
John Moriarty, Engineering Science
Christos Papachristou, Electrical
James E. Wade, Aerospace
Kuo Un Wang, Electrical

1982 - 1983

James M. Briggs, Electrical
David L. Butler,Aerospace
David L. Brown, Mechanical
Michael R. Good, Mechanical
Narayanan Jayaraman, Metallurgical
Peter Kosel, Electrical
Thomas Mantei, Electrical
Robert K. Michael, Engineering Graphics
Anil Mital, Industrial
Prahlad Pant, Civil
Stanley Rubin, Aerospace
Michael G. Ruby, Civil
Frank E. Weisgerber, Civil

1984 - 1985

David Butler, Applied Mechanics
Ernest L. Hall, Mechanical
Herbert Halpern, Mathematics
Sun-Tak Hwang, Chemical
Vikram Kapoor, Electrical
Timothy C. Keener, Civil
Kofi Nyameke, Industrial
Geoffrey Okogbaa, Industrial
Gary Osswald, Aerospace
Sotiris E. Pratsinis, Chemical
Kenneth Roenker, Electrical
Bob W. Stewart, Mechanical

1986

Alireza Behbahani, Nuclear
Pratin Biswas, Civil
Mark T. Bowers, Civil
Danny Cornett, Computer
Peter Disimile, Aerospace
Patrick Eberly, Electrical
Patrick H. Garrett, Electrical
Timothy A. Grogan, Electrical
Howard Jackson, Electrical
Timothy Keener, Civil
Ray Y. Lin, Metallurgical
Lawrence Mazlack, Electrical
Constantine N. Papadakis, Civil
Philip Quo, Mechanical
Panapakkam Ramamoorthy, Electrical
Robert Rost, Mechanical
Debabrata Sarma, Computer
Rodney J. Simmons, Mechanical
Stephen Thiel, Chemical
Raymond Wilkin, Mechanical

1987

Punit Boolchand, Electrical
Linda Cahill, Industrial
Harold W. Carter, Computer
Hong Fan, Electrical
Stevin H. Gehrke, Chemical
Edward S. Grood, Aerospace
Daniel Humpert, Mechanical
Timothy King, Mechanical
Bob Stewart, Mechanical
Bruce K. Walker, Aerospace
Max Wei, Mechanical

1988

Ahmet Aktan, Civil
Paul Bishop, Civil
Robert Brewster, Mechanical
Steven Buchberger, Civil
Kenneth D. Challenger, Metallurgical
Stephen D. Clarson, Materials
Fred B. Cohen, Computer
Atam Dhawan, Electrical
Neville G. Pinto, Chemical
John E. Dickman, Electrical
Altan Ferendeci, Electrical
Frank Gerner, Mechanical
Lee Harrisberger, Mechanical
Thomas R. Hauser, Environmental
Robert G. Jenkins, Chemical
Dale Lawrence, Electrical
Richard A. Miller, Civil
Steven K. Pollack, Materials
J. A. Sekhar, Materials Science
Bahram Shahrooz, Civil
Bruce D. Smith, Electrical
Atmaran H. Soni, Mechanical
Andrew J. Steckl, Electrical
R. Scott Summers, Environmental
Vijay K. Vasudevan, Materials
Philip Wilsey, Computer

1989

Shaaban A. Abdallah, Aerospace
George Batta, Mechanical
Marc M. Cahay, Electrical
Amy R. Ciric, Chemical
Debra A. Hensgen, Computer
Keith D. Keefer, Materials Science
Suk Lee, Mechanical
Glenn G. Lipscomb, Chemical
Allen M. Lush, Aerospace
Sankaran Mohan, Mechanical
Byron P. Newberry, Aerospace
Farhad Radpour, Electrical
Dev Sathyadev, Mechanical
Amir Shirkhodale, Mechanical
George J. Simitses, Aerospace
Makram T. Suidan, Environmental
Ranganadha Vemuri, Computer
Trevor W. Williams, Aerospace

1990

Sundararaman Anand, Mechanical
Jose C. Arantes, Mechanical
Ashraf M. Genaidy, Mechanical
Arthur Helmicki, Electrical
Ali A. Houshmand, Mechanical
Michael J. Kazmierczak, Mechanical
Jonghyuk Kim, Mechanical
Paul D. Orkwis, Aerospace
Raj N. Singh, Materials Science
James G. Uber, Environmental

1991

Albert D. Baker, Computer
Relva C. Buchanan, Materials Science
Milovan S. Beljin, Environmental
Dinesh K. Bhatia, Electrical
Larry P. Cooper, Aerospace
Karen C. Davis, Computer
Ismet Gursul, Mechanical
Jude Iroh, Materials Science
Shafiqul Islam, Environmental
San-Mou Jeng, Aerospace
Jerry Yue-Sheng Lin, Chemical
Raj M. Manglik, Mechanical
A. Dain Samples, Computer
Henry Spitz, Mechanical
Oswald N. C. Uwakweh,
  Materials Science

1992

Perry Alexander, Electrical
Ravi Jain, Research,
  International Engineering
Ravi Kothari, Electrical
Allyn Phillips, Mechanical
Mario Polycarpou, Electrical
Stuart Shelley, Mechanical

The Endowed Chairs and Professorships of the College of Engineering

The L.W. Scott and Ruth Alter Ohio Eminent Scholar Chair in Advanced Manufacturing Systems
Atmaran H. Soni (1987-Present)

The Geier Professors of Engineering Education
Cornelius Wandmacher (1972-1975)
Clyde Melvin Adams (1975-1977)
Kirk Valanis (1979-1984)
Louis Laushey (1984-1986)
Constantine Papadakis (1986-Present)

The Paul E.Geier Professor of Robotics
Ernest L. Hall (1984-Present)

The Carl K. Gieringer Ohio Eminent Scholar Chair in Solid State Microelectronics
Andrew Steckl (1988-Present)

The Bradley Jones Professors of Aerospace Engineering and Engineering Mechanics
Odin Elnan (1973-74)
Robert J. Kroll (1980-84)
David Richardson (1984-86)
George A. Becus (1986-89)
Peter J. Disimile (1989-Present)

The Mitchell P. Kartalia Professor of Electrical and Computer Engineering
Joseph T. Boyd (1984-1990)

The Mitchell P. Kartalia Scholar of Electrical and Computer Engineering
Timothy A. Grogan (1987-90)

The Jacob Lichter Professor of Engineering Construction
John P. Cook (1971-89)
Issam A. Minkarah (1989-Present)

The Peter E. Rentschler Professor of Materials Science
Roy O. McDuffie (1962-67)
Fred E. Westermann (1967-91)

The Herman Schneider Professor of Research
Hans Ernst (1959-79)

The William Thoms Chairs of Civil Engineering
Ward Baldwin (1896-1900)
Harry Thomas Cory (1900-02)
Herman Schneider (1906-29)
Howard B. Luther (1929-51)
Cornelius Wandmacher (1951-58)
Louis Laushey (1958-79)
James McDonough (1979-88)
Paul Bishop (1988-Present)

The Professors Emeriti of the College of Engineering

1935
Herman Schneider,
President Emeritus

1947
Robert Clyde Gowdy,
Dean Emeritus

1949
Saul B. Arenson
Henry Max Goettsch
Clyde William Park

1950
Joseph H. Kindle

1951
Edward A. Eberhardt

1953
Howard Bourne Luther

1954
Earl F. Farnau
Edward S. Smith

1956
Louis Brand

1958
Bartholomew J. Shine

1959
Arthur H. Knebel

1960
Reginald C. McGrane
Harry L. Miller

1961
Reuel Smith

1963
C. Albert Joerger,
Dean Emeritus
David Mazkewitsch
H. Calvert Messinger
Frederick Rogers
Ralph A. Van Wye

1964
Harvey E. Drach
Hans Ernst

1965
Robert E. Hundley

1966
Lewis Franklin Palmer
Raymond W. Renn

1967
Joseph W. Bunting
Howard K. Justice,
Dean Emeritus
Roy O. McDuffie
Meyer Salkover

1968
Herbert A. Dangel
John Warren Surbaugh

1969
Ira David Wollerman

1971
R. Paul Harrington

1973
Bruce B. Baldwin

1975
Arthur E. Focke

1977
John L. Baker
Russell Dunholter
Charles W. Gruber
Albert C. Herweh

1979
Carl F. Evert
Harry Ferguson
William H. Hoblitzell
George Rieveschl, Jr.
Jacob H. Sarver
James Sheng

1980
Edward J. Clearly

1982
Cornelius Wandmacher,
Dean Emeritus

1984
Alexander B. Bereskin
Marvin L. English
Robert J. Fopma
Robert T. Howe
Clair M. Hulley
John N. Pattison

1985
Stanley L. Cosgrove
Robert Lemlich
William Licht

1986
John David Eye
William E. Restemeyer

1987
Louis M. Laushey
Boyd C. Ringo
William Tholke

1988
John P. Cook
Robert M. Delcamp
Richard H. Engelmann
Michael Hoch
Dean Shupe
Francis Tse

1989
James Anno
Roger Messick
James Morand
Robert H. Raible

1990
William Harlow
Daniel Schleef
Widen Tabakoff

1991
William Middendorf
Herbert Preul
Fred Westermann

1992
Donald Hagedorn
James Thorpe
Y.G. Tsuei

Engineering Tribunal Teaching Award

Awarded by the Engineering Tribunal to a member of the college faculty in recognition of excellent teaching.

1969
Robert J. Kroll, Aerospace
B.D. Mullins, Electrical
Carl H. Osterbrock, Electrical
Chris E. Passerello, Engineering Science
Boyd C. Ringo, Civil

1970
Stephen D. Antolovich, Materials Science
Frank E. Burris, Electrical
Robert M. Delcamp, Chemical
William R. Harlow, Engineering Science
Daniel Hershey, Chemical

1971
Max L. Brown, Mechanical
John P. Cook, Civil
Monte J. Pool, Materials Science
Boyd C. Ringo, Civil
Gerald H. Weidenhamer, Aeronautical

1972
Kirti N. Ghia, Aeronautical
Daniel Hershey, Chemical
Ronald L. Huston, Mechanical
Joseph H. Nevin, Electrical
Casimir J. Rosa, Mat. Sci. and Metallurgical

1973
James N. Anno, Nuclear
John David Eye, Environmental
Robert J. Kroll, Aeronautical
James F. McDonough, Civil
Boyd C. Ringo, Civil

1974
William R. Harlow, Engineering Science
Robert J. Kroll, Aeronautical
Monte J. Pool, Materials Science
I-Chich Wang, Mechanical
Michael J. Werle, Aeronautical

1975
James N. Anno, Nuclear
James F. McDonough, Civil
Y.G. Tsuei, Mechanical and Mathematics

1976
William Licht, Chemical
Issam A. Minkarah, Civil
Chris E. Passerello, Engineering Science
Robert H. Raible, Electrical
Dean S. Shupe, Mechanical

1977
Kenneth M. Emmerich, Nuclear
Issam A. Minkarah, Civil
Dean S. Shupe, Mechanical

1978
Eric Bryars, Engineering Science
Robert J. Kroll, Aerospace
Dean S. Shupe, Mechanical

1979
R. Thomas Davis, Aerospace
Alfred W. Scheide, Electrical
William R. Shapton, Mechanical
Richard W. Young, Aerospace

1980
Robert J. Kroll, Aerospace
William Licht, Chemical
Boyd C. Ringo, Civil

1981
Robert M. Delcamp, Chemical
David L. Richardson, Aerospace

1982
Dean S. Shupe, Mechanical

1983
F. James Boerio, Materials Science
Max L. Brown, Mechanical
R. Thomas Davis, Aerospace
Robert Lemlich, Chemical

1984
Kirti N. Ghia, Aerospace
Frank E. Weisgerger, Civil

1985
Thomas D. Mantei, Electrical
Monte J. Pool, Materials Science
Richard W. Young, Aerospace

1986
James M. Morand, Environmental
Joseph H. Nevin, Electrical
Dean S. Shupe, Mechanical

1987
Leroy E. Eckart, Nuclear

1988
Arnold Polak, Aerospace

1989
T. Michael Baseheart, Civil

1990
Robert Ballinger, Mechanical

1991
Arnold Polak, Aerospace

1992
Byron P. Newberry, Aerospace

Outstanding co-op student Vincent A. Rasper, B.S.Ch.E. '66 receives Herman Schneider Medal for exemplary industry-education record from Dean Wandmacher.

Wandmacher Teaching Award

Awarded by the College of Engineering to a member of the faculty in recognition of outstanding teaching performance.

1981 T. Michael Baseheart, Civil
Max L. Brown, Mechanical
Odin Elnan, Aerospace

1982 Kenneth M. Emmerich, Nuclear
Issam A. Minkarah, Civil
Carl H. Osterbrock, Electrical

1983 R. Thomas Davis, Aerospace
James M. Morand, Environmental
Dean S. Shupe, Mechanical

1984 Andrew Bodocsi, Civil
Thomas D. Mantei, Electrical
Monte J. Pool, Materials Science

1985 Leroy E. Eckart, Nuclear
Kirti N. Ghia, Aerospace
James F. McDonough, Civil

1986 Richard W. Young, Aerospace
Robert M. Delcamp, Chemical
Ronald L. Huston, Mechanical

1987 Monte J. Pool, Materials Science

1988 Robert Raible, Electrical

1989 Mark T. Bowers, Civil

1990 Robert J. Kroll, Aerospace

1991 Robert J. Rost, Mechanical

1992 Neville Pinto, Chemical

College Research Award

Awarded to a member of the faculty in recognition of scholarly works and distinguished scientific research.

1987 Joel Weisman, Nuclear

1988 Widen Tabakoff, Aerospace

1989 F. James Boerio, Materials Science

1990 Kirti N. Ghia, Aerospace

1991 Ronald L. Huston, Mechanical

1992 Punit Boolchand, Electrical

Herman Schneider Medal

Awarded to an exemplary co-op student, one who has taken fullest advantage of the unique opportunities of a cooperative engineering education.

1964 Paul W. Marshall, B.S.E.E.
1965 Gary L. Vollbracht, B.S.M.E., and M.S.C.E. '69
1966 Vincent A. Rasper, B.S.Ch.E.
1967 Thomas R. Smith, B.S.E.E., and M.B.A. '68
1968 Lee M. Bauman, B.S.M.E.
1969 Sidney L. Barton, B.S.M.E., andM.B.A. '74
John A. Tomke, B.S.Ch.E.
1970 Gregory A. Graman, B.S.Met.E.
1971 Phillip R. Fairchild, B.S.Ch.E., and M.S.Ch.E. '75
1972 Kenneth C. Hover, B.S.C.E. and M.S.C.E. '74
1973 Roderick D. Davies, B.S.E.E.
1974 James L. Probst, B.S.M.E.
1975 Joseph H. Greenberg, B.S.E.E.
1976 Alan W. Weimer, B.S.Ch.E.
1977 Steven R. Finch, B.S.A.E.
1978 Curtis R. Riffle, B.S.M.E.
1979 Penny L. Bohart, B.S.A.E
Thomas J. Richey, B.S.Ch.E.
1980 Mary J. Ell, B.S.C.E.
Robert R. Ryan, B.S.M.E.
1981 Stephen A. Rothweg, B.S.M.E.
David M. Schuster, B.S.A.E.
1982 Christopher J. Cerimele, B.S.A.E.
1983 Fred W. Obermeier, B.S.E.E.
1984 David B. West, B.S.N.E.
1985 Kevin J. Hemker, B.S.Met.E.
1986 Allen R. Boronkay, B.S.M.E.
1987 Charles M. Royalty, B.S.A.M
1988 Blake A. Sacha, B.S.Ch.E.
1989 Teresa A. Coletta, B.S.M.E.
1990 Greg A. Tipsord, B.S.Ch.E.
1991 Christopher T. Noll, B.S.A.E.
1992 Janet L. Head, B.S.M.E.

Student Organizations of the College of Engineering

Listed in chronological order by either their founding dates at the University of Cincinnati or by their dates of installation as a chapter of a national organization. All descriptions accompany the organizations' first appearances in the Cincinnatian.

1903
Engineers Club
An organization for both scientific and social purposes.

1904
The Sextet
A stringed instrument group from the engineering college.

1908
Co-op Club
A club organized distinctly along social lines, at whose meetings were invited speakers of prominence, from walks of life other than engineering.

Tribunal of the College of Engineering
The judicial body of the College of Engineering.

1910
American Society of Mechanical Engineers
A society to promote the arts and sciences connected with engineering, and to promote better fellowship among the engineering students of the university.

1913
American Institute of Electrical Engineers
Founded with the aim of furthering the cause of the engineer in affording opportunity for the cooperation of the student and men engaged in practical experience. The organization also attempts the dissemination of information as to the development and general trend of the profession of electrical engineering.

1914
Alpha Kappa Psi, Eta Chapter
A professional commercial engineering fraternity.

Braune Civil Engineering Society
Founded as a society to promote a closer relationship among the civil engineers, and to afford opportunity for the presentation and discussion of papers on civil engineering subjects.

Tau Beta Pi, Beta of Ohio
An honorary fraternity for all engineering disciplines.

1917
Alpha Chi Sigma, Alpha Delta Chapter
A professional society for chemistry students in the Colleges of Liberal Arts and Engineering.

1919
Chemical Engineering Club
Founded to promote a stronger feeling of fellowship among the chemical engineers, to further social activities on the part of the chemicals as a group, and to develop the power of student expression on subjects of scientific, technical, and practical importance.

1919
Chemical and Metallurgical Engineers Club
Founded to provide a closer spirit of friendship, to provide a time and place for social gatherings, and to bring the members of the various classes into more intimate contact with one another.

1920
American Society of Civil Engineers
Formerly the Braune Civil Engineering Society.

The Mill-op Club
A club formed by the co-op students of the Cincinnati Milling Machine and Modern Foundry Company for the purposes of more closely uniting the co-ops and their firms.

Women's Co-op Club
The social organization for the co-eps. Later called the Co-ep Club

1921
Co-op Commerce Club
A club founded to foster and inculcate its members a unity of policy and a broader vision concerning industrial and commercial interests.

THE COOPERATIVE ENGINEER
The magazine of the College of Engineering.

Evening Commerce Club
A club founded to promote sociability and to contribute to the welfare of its members and the College of Engineering and Commerce.

Hobby Hour Committee
Founded as the official committee for supervision of hobbies in the college, to pass proposed new hobbies and aid to get them financially on their feet.

Triangle
A social fraternity for engineering and architecture students.

Top: Engineers Club 1907
Middle: Braune Civil Engineering Society 1917
Bottom: American Society of Mechanical Engineers 1912

1922
Beta Gamma Sigma, Alpha of Ohio
An honorary engineering fraternity.

Scroll
An architecture society.

1923
Eta Kappa Nu, Tau Chapter
An honorary electrical engineering fraternity.

Pi Chi Epsilon
Founded as the honorary fraternity of engineering women.

1924
L'Atelier
A professional architectural society.

Delta Sigma Pi, Alpha Theta Chapter
A commercial fraternity.

1925
Delta Mu Delta, Zeta Chapter
An honorary evening commerce fraternity.

Doric Order
Founded as a professional architectural fraternity.

1926
General Engineers Club
A club for students studying general engineering.

American Institute of Chemical Engineers
A professional society of chemical engineers.

Applied Arts Club
A club for the students in the School of Applied Arts.

Pi Tau Sigma, Eta Chapter
An honorary mechanical engineering fraternity.

1927
Men's Tribunal of the School of Applied Arts
The judicial body for men in the School of Applied Arts.

Women's Tribunal of the College of Engineering and Commerce and School of Applied Arts
The judicial body for women.

1928
Co-op Day Committee
The committee of students which plans the events for Co-op Day.

1929
Scarab, Osiris Temple
A professional architectural fraternity.

1930
Women's Tribunal of the School of Applied Arts

1931
Aero Club

APPLIED ARTS JOURNAL
The magazine of the School of Applied Arts.

1932
Chem-Commerce Club

1934
Sigma Kappa Tau
Founded at UC as an honorary applied arts women's fraternity.

1938
Society of Commercial Engineers

1939
Phi Lambda Upsilon, Thirty-Fifth Chapter
Founded to promote high scholarship and to encourage original investigations in all branches of pure and applied chemistry.

1941
Delta Kappa Sigma
An honorary civil engineering fraternity.

Institute of Aeronautical Sciences
Founded to promote fellowship and discussions on aeronautical subjects.

Society of General Engineers

1943
Pi Sigma Pi
Founded as an organization for upperclass general engineering students.

Rocket Society
Founded to further the interest in rocketry, to afford its members the opportunity to increase their knowledge of this new and important field and to do practical and experimental research on the rocket and airstream engines.

1949
Chi Epsilon
A civil engineering honor society.

Top: Co-op Day Committee 1932
Middle: Women's Tribunal 1927
Bottom: Chemical Engineering Club 1920

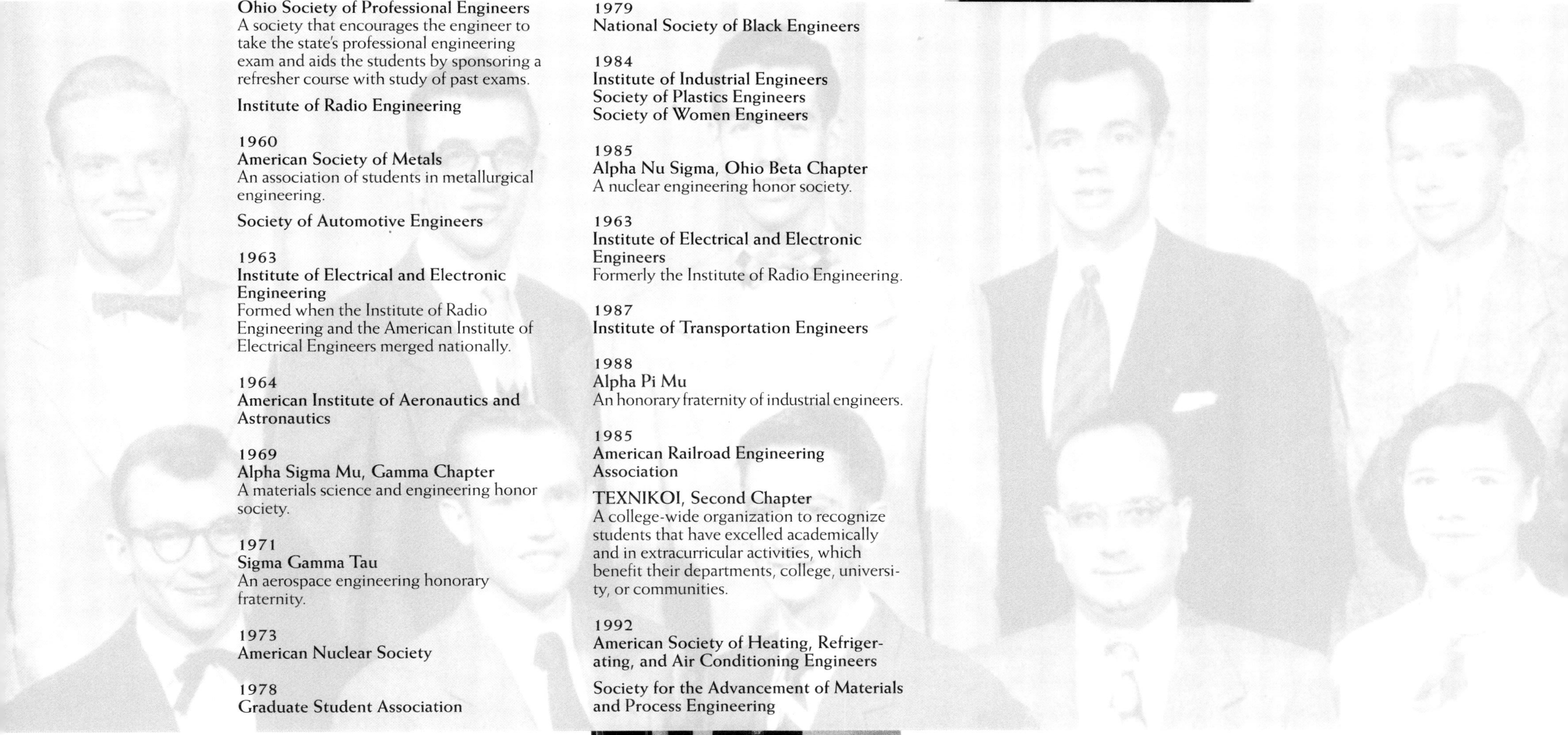

Ohio Society of Professional Engineers
A society that encourages the engineer to take the state's professional engineering exam and aids the students by sponsoring a refresher course with study of past exams.

Institute of Radio Engineering

1960
American Society of Metals
An association of students in metallurgical engineering.

Society of Automotive Engineers

1963
Institute of Electrical and Electronic Engineering
Formed when the Institute of Radio Engineering and the American Institute of Electrical Engineers merged nationally.

1964
American Institute of Aeronautics and Astronautics

1969
Alpha Sigma Mu, Gamma Chapter
A materials science and engineering honor society.

1971
Sigma Gamma Tau
An aerospace engineering honorary fraternity.

1973
American Nuclear Society

1978
Graduate Student Association

1979
National Society of Black Engineers

1984
Institute of Industrial Engineers
Society of Plastics Engineers
Society of Women Engineers

1985
Alpha Nu Sigma, Ohio Beta Chapter
A nuclear engineering honor society.

1963
Institute of Electrical and Electronic Engineers
Formerly the Institute of Radio Engineering.

1987
Institute of Transportation Engineers

1988
Alpha Pi Mu
An honorary fraternity of industrial engineers.

1985
American Railroad Engineering Association

TEXNIKOI, Second Chapter
A college-wide organization to recognize students that have excelled academically and in extracurricular activities, which benefit their departments, college, university, or communities.

1992
American Society of Heating, Refrigerating, and Air Conditioning Engineers

Society for the Advancement of Materials and Process Engineering

Top: Eta Kappa Nu 1969
Middle: The Cooperative Engineer staff 1952
Bottom: ASM 1972

Editors of The Cooperative Engineer

In 1943 The Cooperative Engineer ceased publication for the duration of World War II. It resumed publication in 1946.

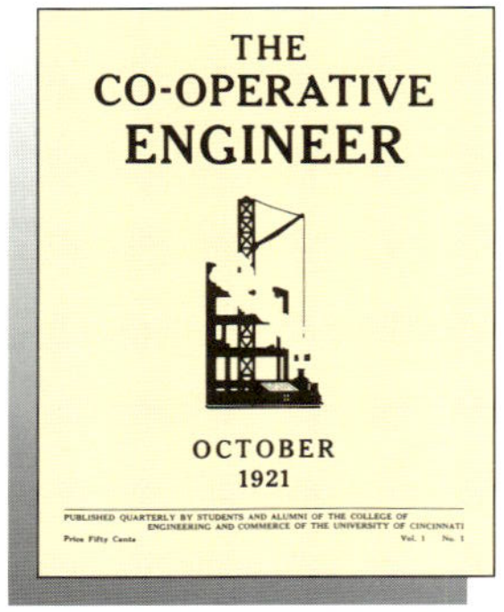

1921

1978

1921 Harry Pockras, C.E. '21

1922 C. Adair Harrell, C.E. '21, D.L. '63

1923 A.F. Knoblaugh, B.Engr. '26

1924 Alexander J. Wildman, Jr., Ch.E., M.S.Ch.E. '26

1925 Harold Senf, C.E. '25
S. J. Miller, Ch.E. '25

1926 Harold C. Hose, C.E. '27

1927 James J. Campbell, Ch.E. '28

1928 Charles T. Hendrick, Engr. attended

1929 Selma F. Hermann, Ch.E. '30

1930 Carl Heyel, E.E. '31

1931 Edwin E. Caspell, M.E. '32

1932 Richard W. Voslamber, M.E. '33

1933 Aaron F. Levy, E.E. '34

1934 Rost E. Warden, Com.E., '35

1935 Charles E. Kaufman, Ch.E. '36

1936 Frank Joseph Messmann, Jr., Ch.E. '37

1937 Donald Montgomery, Ch.E. '39, Ph.D. '45

1938 David Rosenberg, Ch.E. '39

1939 John Meyer, A.A. attended

1940 Douglas Vest, Ch.E. '42

1941 Ruth Mueller, Com.E. '42
Joseph Gump, Ch.E. '43

1942 Edward S. Miller, E.E. '43

1946 John Kemeny, M.E. '47

1947 Walter Friedlander, M.E. '47
Alexander Stolley, M.E. '48

1948 Rober Laib, Ch.E. '49

1949 Alexander Batscha, Ch.E. '50

1950 Robert F. Hausman, Ch.E. '51

1951 Park Gast, Jr., M.E. '52

1952 Raymond Ogle, B.S.Arch. '55

1953 Donald Ebel, Ch.E. '54

1954 Ronald E. Rosenweig, '55

1955 Dick Folkerth, A.E. '57

1956 Ronald Eckelmann, E.E. '58

1957 James Wilger, A.E. '58

1958 John Lower, E.E. '60, M.S.E.E. '62

1959 Bill Black, Met.E. '60

1960 Clifford Buchman, E.E. '61

1961 Tom Myers, B.S.C.E. '67, B.S.M.E. '75

1962 Nash McCauley, B.S.M.E. '64

1963 Tom Driscoll, B.S.Ch.E. '64

1964 Walter Stanley McCauley, B.S.M.E. '65

1965 Ron Lang, B.S.M.E. '66

1966 Jay Kenneth Hay B.S.M.E. '68, M.S.M.E. '72

1967 & 1968 Stanton A. Glantz, B.S.A.E. '69

1969 Thomas Geers, B.S.E.E. '70

1970 & 1971 Chuck Naegeli, B.S.A.E. '71

1972 & 1973 Doug Melhorn, B.S.E.E. '73

1974 Brent Stiles, B.S.Ch.E. '75

1975 Brent Stiles, B.S.Ch.E. '75
Shayne Manning, B.S.C.E. '77

1976 Pat Zavist, B.S.Ind.Mgmt. '81
Terry Brennan, B.S.Ch.E. '80
Shayne Manning, B.S.C.E. '77

1977 Steven Gorsler, B.S.Ch.E. '80
Terry Brennan, B.S.Ch.E. '80

1978 Matthew McGonegle, A.E. '81

Distinguished Alumni of the College of Engineering

The Distinguished Alumnus Award was initiated during the Sesquicentennial of the University of Cincinnati, to recognize outstanding alumni who, by their achievements and stature in engineering, science, industry, business, public service, or education, have reflected an unusually large measure of credit on the University of Cincinnati and the College of Engineering.

1969

Russell L. Ahlbrand, E.E. '48
James E. Allan, C.E. '19
W. D. Allardice, C.E. '24
William M. Allin, Com.E. '33
Charles Arnold, M.E. '23
J. H. Ashley, Ch.E. '49
Ralph H. Auch, Ch.E. '17
Howard E. Barnett, E.E. '30
Leon Bass, E.E. '31
Benjamin B. Bauer, E.E. '37
George W. Binns, M.E. '11
Arthur D. Bird, C.E. '33
Raymond L. Bisplinghoff, A.E. '40
Alfred T. Blackburn, M.E. '30
Horace M. Blinn, Engr. '24
Ralph Bloom, Jr., Ch.E. '41
Robert J. Blum, Engr.Sci. '24
Hector P. Boncher, Com.E. '28
H. E. Bonham, C.E. '28
Spencer B. Booz, M.E. '34
Charles R. Borders, E.E. '43
Lester L. Bosch, E.E. '25
Leonard D. Bradt, Ch.E. '33
Louis Brand, Ch.E. '07
Oscar S. Bray, C.E. '32
G. W. Breiel, C.E. '21
James A. Broadston, M.E. '34
Carlton E. Brown, M.E. '23
L. Lee Burke, A.E. '39
Myrtice C. Burr, E.E. '28
James L. Burridge, A.E. '43
Erskine Chapman, C.E. '42
Frank G. Comerford, C.E. '32
James F. Connaughton, Com.E. '37
Charles Constable, M.E. '21
N. J. Constantine, M.E. '52
Clifford D. Coy, E.E. '29
P. Willard Crane, Ch.E. '29
J. Russ Daniell, A.E. '36
Joseph Dave, C.E. '23
Harold E. Deardorff, E.E. '22
Homer R. Denius, E.E. '41
Robert J. Dineen, E.E. '52
Thomas L. Dineen, E.E. '52
W. J. Distler, M.E. '33
M. R. Dodson, Com.E. '32
John Eastes, Ph.D. '36
J. E. Eberhardt, Ch.E. '33
Don A. Ellis, Ch.E. '41
Albert E. Ernst, C.E. '23
Elliott M. Estes, M.E. '40
Jacob Feld, C.E. '21
Michael Field, Ph.D. '48
Samuel T. Fife, E.E. '26
Dean P. Fite, Com.E. '35
Dean Earl H. Flath, E.E. '19
Harold L. Flowers, M.S.E.E. '48
Peter L. Frommer, E.E. '54
C. L. Gallimore, C.E. '28
Harold J. Garber, Ch.E. '35
James H. Garrison, Ch.E. '20
Stuart Garrison, Engr.Sci. '22
Bruce M. Geiger, M.E. '42
Carl K. Gieringer, E.E. '32
Robert S. Goebel, M.E. '37
Herbert J. Goosman, M.E. '16
Arthur C. Greber, Ch.E. '38
Robert S. Hahn, M.E. '40
W. A. Hallett, M.E. '25
Hunter W. Hanly, M.E. '06
J. F. Hansjergen, Ch.E. '44
C. A. Harrell, C.E. '24
J. Reed Hartman, Ch.E. '20
Chester J. Henry, C.E. '23
Mark P. Herschede, E.E. '40
Jack W. Herbstreit, E.E. '39
Russell Hoffman, Met.E. '47
J. Benjamin Horvay, E.E. '43
H. C. Hose, C.E. '27
A. O. Hurxthal, M.E. '14
Dorman D. Israel, E.E. '23
William J. Ivers, C.E. '31
David W. James, M.E. '42
Richard Jarrold, Ch.E. '42
William H. Jefferson, E.E. '12
Donn O. Jennings, Com.E. '41
Hal J. Jones, M.S. '47
John F. Jordan, E.E. '32
M. P. Kartalia, E.E. '40
Robert J. Kerr, Ch.E. '40
John E. Keto, E.E. '32
Robert B. Killen, E.E. '36
Thomas T. Kling, M.E. '33
Walter E. Knocke, Com.E. '38
Winston E. Kock, E.E. '32
Christopher L. Koehler, M.E. '19
Robert M. Koppenhoefer, Ch.E. '33
Emil J. Kratt, C.E. '23
John W. Krausser, C.E. '25
John W. Knueger, M.E. '30
Ralph H. Knuse, E.E. '16
Leo G. Kuhlman, E.E. '17
Ralph L. Langenheim, C.E. '17
Howard Lepple, E.E. '34
R. E. Lippert, M.E.'47
Walter E. Littmann, Met.E. '50
Charles R. Lowry, Ch.E. '41
Hugh D. Luke, Ch.E. '34
Matthias E. Lukens, Com.E. '35
Charles F. MacNish, C.E. '30
W. O. Mashburn, Jr., Com.E. '30
Howard A. McGlasson, A.E. '38
John W. Melville, M.E. '32
M.Eugene Merchant, M.S.Engr. '38, Sc.D. '41
Louis J. Messer, C.E. '37
Frank J. Messmann, Ch.E. '37
Samuel J. Miller, Ch.E. '25
Don G. Mitchell, Engr. '26
Harry W. Moeller, Ch.E. '39
Donald J. Montgomery, Ch.E. '39, Ph.D. '45
R. Dail Moore, E.E. '36
Lester C. Morrow, M.E. '12
George H. Muth, Com.E. '23
James H. "Herb" Myers, M.E. '29
Robert W. Neel, Com.E. '35
William R. Oeh, Ch.E. '40
Theodore F. Olt, Ch.E. '28
Cyrus R. Osborn, M.E. '21
Edward C. Pandorf, Com.E. '40
Wilbur J. Peets, M.E. '13
James C. Perin, C.E. '35
John B. Petzhold, Jr., Com.E. '23
Phillip K. Pfanner, E.E. '33
Harry Pockras, C.E. '21
Edwin F. Price, Met.E. '47
Robert A. Ramey, E.E. '43
William E. Ranz, Ch.E. '47
Thomas C. Raudebaugh, A.E. '40
J. K. Records, E.E. '48
Oscar B. Reemelin, C.E. '04
Charles M. Reesey, M.E. '28
R. E. Reimer, Com.E. '28
Frank S. Riordan, Jr., Ch.E. '40
Max B. Robinson, M.E. '12
C. F. Roby, M.E. '22
H. Alex Romanowitz, E.E. '24
Edward L. Roth, Met.E. '24
C. Oscar Schmidt, M.E. '31
Frank W. Schroeder, C.E. '30
Herbert H. Schroth, C.E. '16
A. W. Schubert, Com.E. '24
Philip W. Scott, Com.E. '33
Milton C. Shaw, Sc.D. '42
Charles S. Shryock, E.E. '19
Kelly Y. Siddall, Com.E. '25
Edwin J. Smith, Ch.E. '40
Howard Snively, E.E. '34
George Sperti, E.E. '23
E. E. Stansbury, Ph.D. '46
Edwin P. Stenger, Met.E. '16
Earl G. Stegemiller, Com.E. '33
John L. Strubbe, Engr. '47
James M. Stuart, Engr. '26
Carl F. Stugard, M.E. '31
L. E. Swabb, Ch.E.'48
Herbert W. Starick, C.E. '31
Charles W. Taylor, Com.E. '23
Robert R. Templeton, A.E. '35
James G. Terrill, Jr., C.E. '37
Burson Treadwell, M.E. '30
Frank M. Tiller, Ph.D. '46
H. Cal Uihlein, M.E. '19
Richard Van Veen, C.E. '38
Curtis H. Vaughan, M.E. '34
Rudy B. Vogt, M.E. '25
James D. Voss, Met.E. '50
Berz S. Wagner, E.E. '22
James P. Walsh, Jr., Com.E. '33
Calvin J. Werner, E.E. '30
Carson R. Whiting, Com.E. '33, D.C.S. '84
W. Harmon Wilson, Com.E. '29
Raymond L.Woerner, Com.E. '35
Ballard A. Yates, Ch.E. '29
B. John Yeager, E.E. '34
Everett B. Yelton, Ch.E. '40
William H. Zimmer, Engr. '25

1970

Robert Boni, Met.E. '51
Oscar W. Dillon, A.E. '51
H. L. Eiselstein, Met.E. '41
Richard J. Francis, Ch.E. '34
Ben R. Haverstick, M.E. '33
Gordon A. Hughmark, Ch.E. '48
James F. Kaiser, E.E. '52
George J. Kral, C.E. '43
Robert C. Lewis, C.E. '32
Martin Littmann, Met.E. '41
Leigh M. Matthews, E.E. '43

Distinguished Young Engineer Executive

Awarded to alumni for achievement and potential for conspicuous success in industry.

1987
David R. Levering,
B.S.M.E. '66
General Motors Corporation

1988
Roderick D. Davies,
B.S.E.E. '73
Texas Instruments, Inc.

1989
Ernesto Marco Aieta,
B.S.C.E. '78
James M. Montgomery,
Consulting Engineers, Inc.

1990
Albert E. Vierling,
B.S.E.E. '76
Computer Technical
Corporation

1991
Alan W. Weimer,
B.S.Ch.E. '76
Dow Chemical Corporation

1992
William D. Miller,
B.S.Eng.Sci. '81
Digital Communications
Associates, Inc.

Louis Schriber, M.E. '35
1971
Robert M. Goldhoff, Ch.E. '43
Charles E. Huckaba, Ph.D. '53
Warner L. Stewart, A.E. '49
Charles V. Theis, C.E. '22
1972
John C. Chato, M.E. '54
Thomas D. Cooper, Met.E. '55
James T. Donahue, E.E. '52
Edward L. Fleckenstein, E.E. '38
John F. Kahles, Ch.E. '36
Paul L. Pfennigwerth, M.E. '52
Leo L. Weber, E.E. '53
1973
J. Floyd Byrd, Ch.E. '48
Walter Cornelius, A.E. '38
Carson O. Donley, M.E. '39
W. Benjamin Evans II, M.E. '53
James E. Funk, C.E. '55
Howard K. Justice, C.E. '19
Willard E. Kemp, M.E. '55
Robert E. Parsons, M.E. '54
Edgar E. Parks, C.E. '51
Scott H. Simpkinson, M.E. '43
Thomas E. Smith III, Met.E. '48
Edward F. Weller, E.E. '43
1974
Arthur R.Ehrnschwender, M.E. '48,
B.S.Com. '59
C. Edward Glass, M.E. '36
Ammand F. Knoblaugh, Engr. '25
Paul G. Nelson, Ch.E. '40
William F. Richards, C.E. '41
C. Martin Stickley, E.E. '57
James R. Weir, Jr., Met.E. '55
Lloyd M. Whitesell, M.E. '42
1975
Linton D. Bylund, C.E. '52
P. Kenneth Seidelmann, E.E. '60,
M.S. '62, Ph.D. '68
James H. Smith, E.E., M.S.Engr. '31
James A. Theys, M.E. '49
1976
Alfred C. Berndsen, C.E. '52
Harry L. Bishop, Jr., Met.E. '51
Allen W. Brown, Ch.E. '55
Donald E. Haok, A.E. '54
Richard C. Messinger, M.E. '53

Judith A. Schwan, Ch.E. '48
Charles J. Sheridan, E.E. '43
Glenn L. Smith, M.E. '52
1977
Ralph C. Bursiek, Com.E. '31
George Fair, A.E. '57
William P. Koster, M.S. '31, Ph.D. '53
Boise F. Macon, C.E. '43
Alan L. Myers, Ch.E. '60
1978
Carl J. Altstetter, Met.E. '53
Ralph S. Binns, Ch.E. '43
Russell Dunholter, C.E. '30
John H. Hall, E.E. '61
W. Leonard Harper, C.E. '28
Albert C. Herweh, E.E. '32, M.S.E.E. '36
Jeffrey C. Kalb, E.E. '65
Jason R. Lemon, M.E. '58
Paul G. Mayer, C.E. '53
J. Frank Shawhan, E.E. '35
William L. Taylor, Ch.E. '54
1979
Richard D. Anderegg, C.E. '54
O. Lester Applegate, D.C.S. '71
Lynn E. Arnold, Met.E. '57
John D. Breslin, Ch.E. '55
Robert F. Denig, C.E. '41
Walter E. Foster, Ch.E. '49
Herbert M. Fuldner, E.E. '54
Clyde G. Haehnle, E.E. '44
F. Robert Heyner, A.E. '53
Herbert M. Katz, Ph.D. '54
1980
Lamar Cox, E.E. '56
William H. Dickhoner, M.E. '50
Charles W. Gruber, M.E. '32
William R. Martin, Met.E. '57
Charles S. Meyer, E.E. '59
John R. Miller, Ch.E. '60
Earl A. Seamands, C.E. '16
D. S. Seth, M.S.Ch.E. '56
Merlin G. Smith, E.E. '50
1981
Clark E. Beck, M.E. '55, M.S.A.E. '69
John Brombaugh, E.E. '60
Arthur L. Fricke, Ch.E., '57
Jerome B. Gilbert, C.E. '53
Louis F. Kazda, E.E. '40, M.S.E.E. '43
Dale Martin Kohler, Met.E. '56

James H. Leonard, Ch.E. '55
Juris Vikmanis, E.E. '61
William Lawrence Wachs, E.E. '53
1982
R. Gerald Colclaser, E.E. '56
John B. Guernsey, Met.E. '52
James R. Krusling, C.E. '48
Robert A. Scholtz, E.E. '58
Louis J. Urban, E.E. '56
1983
Henry Thomas Brown, Ch.E. '55
James E. Kroencke, M.E. '59
Lawrence A. Lantzer, A.E. '61
Charles J. Marshall, E.E. '39
Paul W. Marshall, B.S.E.E. '64
Dale T. Peters, Met.E. '58
Thomas E. Petry, M.E. '62
1984
Jack N. Binns, Sr., M.E. '40
Donald L. Birkimer, M.S.C.E. '65,
Ph.D. '68
C. Thomas Cross, Ch.E. '50
Elihu Geer, C.E. '24
J. Robert Jump, E.E. '60, M.S.E.E. '62
Denis J. Prager, E.E. '62
Richard J. Schaffnit, Met.E. '58
Edwin A. Stenger, Met.E. '51
S. Jay Stewart, Ch.E. '61
Donald L. Vieth, Met.E., '63, Ph.D.'72
Robert C. Williams, M.E. '53
Robert W. Wiwi, B.S.E.E. '64
Linwood C. Wright, A.E. '44,
M.S.A.E. '60
1985
Alexander B. Bereskin, E.E. '35,
M.S.E.E. '41
Frank P. Fazzari, Met.E. '55
William G.H. Finch, E.E. attended
Barry W. Hannah, A.E. '63, M.S. '65,
Ph.D. '73
Melville D. Hensey, C.E. '57, M.S.C.E. '68
John Paul Hyde, E.E. '57
Edwin L. Mongan, Ch.E. '62
James R. Myers, Met.E. '56
Donald Smally, M.E. '49
1986
Arthur A. Astbury, Met.E. '63
William C. Boyle, C.E. '59, M.S. '60
Odin Elnan, B.S.A.E. '51, M.S.A.E. '56

George K. Grant, Ph.D. '73
Robert Theo Howe, C.E. '43, M.A. '55
William Licht, Ch.E. '37, M.S.Ch.E. '39, Ph.D. '50
John G. Proakis, E.E. '59
Ronald E. Rosenswig, Ch.E. '55
Alfred W. Scheide, B.S.E.E. '62,
M.S.Ch.E. '64, Ph.D. '72
Paul G. Schloemer, M.E. '51
1987
Julio R. Bartol, Met.E. '62
Leonard Beitch, M.S. '59, Ph.D. '68
John J. Coy, M.E. '65, M.S.M.E. '69, Ph.D. 71
William J. Kramer, Jr., M.E. '63, M.S.M.E. '67
James A. Maish, Ch.E. '54
Perry P. Yaney, E.E. '54, M.S. '57, Ph.D. '63
1988
Rodney A. Boudreau, Ph.D. '65
James R. Hyde, Ch.E. '62
Albert L. Klosterman, B.S.M.E.'65,
M.S.M.E. '68, Ph.D. '71
Ronald F. Layton, M.S. '68, Ph.D. '70
Dane A. Miller, M.S.Met.E. '71, Ph.D. '74
Roderick D. Davies, B.S.E.E. '73
1989
Russel F. Henke, M.E. '63, M.S. '65, Ph.D. '68
Joseph A. Mancini, A.E. '35
Norman E. Nelson, A.E. '41
Howard W. Pickering, Met.E. '58
Roger H. Zanitsch, B.S.C.E. '67, M.S.C.E. '69
1990
Frank G. Jones, Ch.E. '37
William D. Lockwood, C.E. '57, M.S.C.E. '69
Norman D. Malmuth, A.E. '53
Vinod Kumar Sikka, M.S.Engr.Sci. '70, Ph.D.'73
Thomas A. Zimmerman, B.S.E.E. '65
1991
Eugene R. Allspach, B.S.Ch.E. '70
Donald J. Dusa, M.S.A.E. '62
Ayhan Hakimolgu, M.S.E.E. '50
Darrell F. Socie, B.S.Met.E. '71, M.S.Met.E.'73
Peter S. Strange, B.S.C.E. '72
Chih-han Wang, M.S.M.E. '66, Ph.D. '74
1992
Harry Hollack, B.S.E.E. '67
James P. McGlone, M.E. '54
William D. Miller, B.S.Engr.Sci. '81
Vikram Rajadhyaksha, M.S.C.E. '69
Gregory M. Reck, B.S.A.E. '69
John Tomke, B.S.Ch.E. '69

College of Engineering Alumni Board Members

The Alumni Board was founded in 1973 to promote the interests of the college and enhances the relationship between the College of Engineering and its alumni. Date indicates first year of term.

1992	Eugene R. Allspach (B.S.Ch.E. '70)
1977	James Armour (B.S.Ch.E, '61)
1975	Lynn Arnold (B.S.Met.E. '57)
1978	Paul Gary Ater (C.E. '63)
1988	Clark E. Beck (M.E. '55, M.S.M.E. '69)
1992	Jon A. Bennett (M.S.C.E. '73)
1977	Edward Buynacek (M.E. '57)
1988	John T. Cammett, III (Ph.D. '73)
1988	David L. Chalk (Ch.E. '62)
1982	James C. Champlin, III (B.S.E.E. '77)
1988	Steven Crowley (B.S.M.E. '83, M.S.M.E. '84)
1979	Kenneth Deem (Ch.E. '62)
1992	Mina Dimov (B.S.M.E. '83, M.S.M.E. '87)
1988	Richard E., Dippery, Jr. (B.S.M.E. '65, M.S.M.E. '71, Ph.D'88)
1980	Jack E. Drake (M.E. '52)
1975	George C. Dreyer (E.E. '57)
1988	James Dunn (B.S. '79, M.S.M.E. '82)
1975	Arthur Ehrnschwender (M.E. '48, B.S.Com. '59)
1982	James J. Flaig (C.E. '60, M.S.C.E. '68)
1974	William Forbriger (Ch.E. '53)
1982	Luciano R. Gatto (B.S.M.E. '62)
1975	James T. Gavin (B.S.M.E. '56)
1978	Edward Glass (M.E. '36)
1979	Franklin Grooms (B.S.E.E. '55)
1974	Robert G. Hendon (B.S.M.E. '65)
1975	Melville D. Hensey, III (C.E. '57, M.S.C.E. '68)
1978	Edward Hirsch (M.E. '44)
1978	Robert Kispert (Ch.E. '60, M.S.Ch.E. '63)
1988	Richard D. Koch (B.S.M.E. '70, M.B.A. '75)
1977	John Michael Laumann (B.S.C.E. '73, J.D. '77)
1975	Richard Lenz (B.S.C.E. '66, M.S. '68)
1992	William E. Lower (M.E. '63, M.S.M.E. '65)
1984	Daniel W. McCurdy (B.S.Met.E. '77)
1984	Orest Melnyk (B.S.E.E. '80, M.B.A. '86)
1976	Richard C. Messinger (M.E. '53)
1981	Francis (Gene) Moenck (B.S.C.E. '58)
1979	W. Bruce Motz (M.E. '51)
1973	Richard D. Neidhard (M.E. '51)
1988	Janet Chiason Nickloy (B.S.M.E. '84, M.S.M.E. '87)
1983	Howard W. Renner (M.E. '59)
1983	Sidney G. Rindsberg (Ch.E. '43)
1974	Gerard Roberto (C.E. '55)
1992	Mark Sager (M.S.M.E. '90)
1992	Ronald Schallick (B.S.M.E. '70, M.B.A '73)
1988	Alfred Scheide (E.E. '62, M.S.Ch.E. '64, Ph.D. '72)
1975	James Schoelwer (C.E. '55)
1984	Richard L. Schwettman (B.S.M.E. '75)
1975	Joseph J. Seiwert (Ch.E. '59)
1980	Gerald L. Shawhan (Met.E. '58, Ph.D'66)
1988	Charles J. Sheridan (E.E. '43)
1975	Edwin L. Slone (M.E. '51, As.E.E. '59)
1992	Jeff Smith (E.E. '78)
1982	David Suprock (B.S.Met.E. '74)
1975	William L. Wachs (E.E. '53, M.E. '61)
1984	Jonathan T. Weber (M.E. '61)
1988	Robert P. Wiwi (B.S.E.E. '64)

College Advisory Council

Established in 1988 to help advise the college on long-range needs of industry by identifying technical, research, and business trends.

Name	Company	First Year of Term
Robert Barthelmy	Wright Patterson Air Force Base	1990
Barry Bebb	Xerox Corporation	1988
Alfred Berndsen, C.E. '52	Frank Messer & Sons	1988
Robert Boni, M.E. '51	Armco, Incorporated	1989
Thomas J. Canavan	NEC America	1988
Darrel Clark, B.S.E.E. '64	NCR Corporation	1991
John Cooper, Ph.D. '57	DuPont Company	1990
Herbert Cotter	IBM Corporation	1990
William Dickhoner, M.E. '50	Cincinnati Gas & Electric Company	1988
Thomas Donohue, B.S.C.E. '75	General Electric Company	1988
Philip Francis	Square D Company	1992
Ronald Friedsam	Structural Dynamics Research Corporation	1992
James Geier, D.C.S. '83	Cincinnati Milacron, Incorporated	1990
Richard Glover	Procter & Gamble Company	1988
Charles Joiner	Mead Corporation	1988
Lamar Kerr, M.E. '63	General Motors Corporation	1990
John M. Klineberg	NASA Goddard	1989
John LaMacchia	Cincinnati Bell Incorporated	1989
Richard Lenz, C.E. '66, M. S. '68	Alumni Board Liasion	1990
Dan McGlaughlin, M.E. '59	G.E. Calma	1988
Richard Messinger, M.E. '53	Cincinnati Milacron, Incorporated	1988
Dane Miller, M.S.Met.E. '71, Ph.D. '74	Biomet Incorporated	1990
Frank Mosier	BP America	1990
Robert Murphy	James River Corporation	1988
Vikram Rajadhyaksha, M.S.C.E. '69	Dodson-Lindholm Associates	1991
Keith Richey	Wright Patterson Air Force Base	1992
Raymond Ross, As. '65, Bach. '68	Cincinnati Milacron	1992
George Schaefer	Fifth Third Bank	1989
Paul Schloemer, M.E. '51	Parker Hannifin Corporation	1988
Charles Sheridan, E.E. '43	Alumni Board Liaison	1988
S. Jay Stewart, Ch.E. '61	Morton International	1988
Thomas Stokes, Bus. '58	General Motors Corporation	1988
George Strike	American Laundry Machinery	1988
John Tomke, Ch.E. '69	Dow Chemical Company	1988
James Van Tassel	NCR Corporation	1990
Juris Vikmanis	Square D Company	1989
Robert Wiwi, B.S.E.E. '64	Alumni Board Liaison	1992
Ronald H. Yocum	Quantum Chemical Corporation	1989
Roger Zanitsch	Calgon Carbon	1990

Key to Degree Abbreviations

The following degree abbreviations appear throughout the text.

A.A. - baccalaureate degree in applied arts, from the College of Engineering and Commerce and School of Applied Art

A.E. - baccalaureate degree in aeronautical engineering

As. - associate's degree, earned from either the Evening College or the College of Applied Science

B.A. - bachelor of arts, earned from the College of Arts and Sciences

B.S. - bachelor of science, earned from either the College of Engineering or the College of Arts and Sciences

B.S.Arch. - bachelor of architecture, earned from the College of Engineering prior to 1946 and from the College of Applied Arts after 1946

B.S.Com. - bachelor of science in commerce, earned from the College of Business Administration

B.S.A.E. - bachelor of science in aeronautical engineering prior to 1974, bachelor of science in aerospace engineering after 1974

B.S.A.M. - bachelor of science in applied mechanics

B.S.C.E. - bachelor of science in civil engineering

B.S.Ch.E. - bachelor of science in chemical engineering

B.S.E.E. - bachelor of science in electrical engineering

B.S.Engr.Sci. - bachelor of science in engineering science

B.S.I.E. - bachelor of science in industrial engineering

B.S.Ind.Mgmt. - Bachelor of Science in Industrial Management, earned jointly from the College of Engineering and the College of Business Administration

B.S.M.E. - bachelor of science in mechanical engineering

B.S.Met.E. - bachelor of science in metallurgical engineering

B.S.N.E. - bachelor of science in nuclear engineering

Bus.Ad. - baccalaureate degree in business administration, earned from the College of Engineering and Commerce prior to 1946 and from the College of Business Administration after 1946

C.E. - baccalaureate degree in civil engineering

Cert. - certificate, earned from the Evening College or the College of Applied Science

Ch.E. - baccalaureate degree in chemical engineering

Com.E. - baccalaureate degree in commercial engineering

D.C.S. - honorary degree of doctor of commercial science

D.F.A. - honorary degree of doctor of fine arts

D.L. - honorary degree of doctor of laws

D.Sc. - honorary degree of doctor of science

E.E. - baccalaureate degree in electrical engineering

Engr. - baccalaureate degree in general engineering

Engr.Sci. - baccalaureate degree in engineering science

I.E. - baccalaureate degree in industrial engineering

M.B.A. - master of business administration, earned from the College of Business Administration

M.A. - master of arts, earned from the Graduate School in a non-engineering program

M.E. - baccalaureate degree in mechanical engineering

Met.E. - baccalaureate degree in metallurgical engineering

M.S. - master of science, earned from the Graduate School in a non-engineering program

M.S.C.E. - master of science in civil engineering

M.S.Ch.E. - master of science in chemical engineering

M.S.E.E. - master of science in electrical engineering

M.S.Engr. - master of science in general engineering

M.S.M.E. - master of science in mechanical engineering

M.S.Met.E. - master of science in metallurgical engineering

Ph.D. - doctor of philosophy, earned from either the College of Engineering or the College of Arts and Sciences

Sc.D. - doctoral degree in science, earned from the Institute of Scientific Research

Selected Bibliography

The majority of references for this book are historical documents which were compiled, filed, and maintained by Cornelius Wandmacher throughout his career as the William Thoms Professor, dean, and dean emeritus of the college. They are housed in the Archives and Rare Books Department at the University of Cincinnati.

Adams, Charles Francis, ed. *The Memoirs of John Quincy Adams*. Vol. 11, *Comprising Portions of His Diary from 1795 to 1848*. Philadelphia: J. B. Lippincott & Co., 1876.

Bauer, William Charles. "The Lion-Hearted Class of 1896." *Cincinnati Alumnus* 34, no. 5 (July 1960): 8-10.

Billinger, R. D. "A Night in Alchemy." *Journal of Chemical Education* 5, no. 6 (June 1928): 715-24.

Brannon, Joseph A. "A Biographical Sketch of Richard J. Francis." Prepared for Clara Francis. Newark, Ohio, 1986.

Burns, Paul. "Where All Branches of Science are Wed." *Cincinnati Alumnus* 7, no. 2 (November 1934): 6.

"Cincinnati Facility is Selected 'Lab of the Year.'" *Industrial Research* 13, no. 5 (May 1971): 21-22.

The Cooperative Engineer 1, no. 1 to 53, no. 4 (October 1921-January 1978).

Curran, Chris. "Fax-ination: Telecommunications Pioneer Gives the Facts about Fax." *UC This Week* 14, no. 33 (8 June 1990): 4.

The Debris. Vol. 36 and Vol. 50 West Lafayette: Purdue University, 1924 and 1938.

Golden, Dixie W., ed. *A Golden Gate Bridge Jubilee, 1937-1987*. Cincinnati: University Publications, 1987.

Heidenreich, Rob. "Engineering College Gets Interim Leader." *The News Record* 70, no. 26 (9 March 1983) 1.

Hulley, Clair. "More Memories of Fire." *UC Currents* 1, no. 9 (22 November 1991): 2.

Hurter, Jerry. "Stars of 1915 Bearcat Team Reminisce." *Hyde Park Living* 10, no. 12 (December 1991) 10-11.

"Innovative Programs Enhance Student Opportunities." *UC Engineer* 3, no. 1 (Spring 1991): 4-5.

Laketown, Jack. "A New Science of Light is Born in an Attic Laboratory." *Cincinnati Alumnus* 3, no. 3 (February 1931): 5-7.

Langsam, Walter C. *At the University of Cincinnati, 1955-1971: A Memoir*. Cincinnati: The University of Cincinnati Foundation, 1982.

McGrane, Reginald C. *The University of Cincinnati: A Success Story in Urban Higher Education*. New York: Harper & Row, 1963.

McKay, Robert. "The Amazing Dr. Sperti." *Cincinnati Magazine* 15, no. 11 (August 1982): 46-52.

Mitchel, F. A. *Ormsby MacNight Mitchel, Astronomer and General*. Boston: Houghton Mifflin, 1887.

"On the Campus." *Cincinnati Alumnus* 17, no. 2 (Winter 1944-45): 2.

Park, Clyde W. *Ambassador to Industry: The Idea and Life of Herman Schneider*. Indianapolis: Bobbs-Merrill, 1943.

"Pinhead Electronics." *UC Alumnus* 41, no. 5 (April 1967): 15.

Rieselman, Deborah. "When UC Went to War." *Horizons* 21, no. 2 (October 1991): 5-20.

Schabo, Andi. "Pioneer Got His Fax Right." *The Port St. Lucie News*, 53 no. 349 (17 June 1990): 1.

Schneider, Herman. "Technical Education for Cincinnati." *American Machinist* 28 (19 October 1905): 540-41.

"Science World Turns Eyes To Garret Of University's Basic Science Laboratory, Whence Came Astounding Food Discovery." *Cincinnati Enquirer* 16 January 1931.

"A Timely Return: UC's Sun-Clock is Back on Its Pedestal." *UC Alumnus* 45, no. 1 (October 1970): 9.

"UC Instrumental in Developing Ohio Aerospace Institute." *UC Engineer* 2, no. 1 (Winter 1990): 2-3.

"The University and Defense." *Cincinnati Alumnus* 13, no. 1 (Fall 1940): 2.

"W8CAU." *QST* 13, no. 11 (November 1929): 34-38, 84-86.

Walters, Raymond. *The Centennary of the Cincinnati Observatory*. Cincinnati: Court Index Press, 1944.

———. *Historical Sketch of the University of Cincinnati*. Cincinnati: The Mountel Press, 1940.

Photograph Credits

All photographs are the property of the University of Cincinnati Archives and Rare Books Department, with the following exceptions:

The photographs on the following pages are the property of the University of Cincinnati College of Engineering: 19, 23, 24, 25, 26, 27, 30, 34, 35, 46 (William Osterbrock), 48, 53 (faculty), 54 (Schneider Foundation), 56 (College of Engineering staff), 58, 60, 62, 63, 64, 67, 69, 71, 73, 74, 75, 77, 78, 79 (C. Melvin Adams), 82 (laboratory and Kirk Valanis), 83, 84, 106 (Herman Schneider), 107 (Louis Laushey and Constantine Papadakis), and 117.

The photographs on the following pages are reprinted from *The Cooperative Engineer*: 22, 37, 38, 40, 49, and 72 (illustration).

The photographs on the following pages are reprinted from the University of Cincinnati's annual, *The Cincinnatian*: 21 (Charles Dabney), 42, 43, 52 (statistics laboratory), 68, 92 (UC Bearcats), 106 (Thomas Evans), 118, 119, 120, and 121.

The photographs on the following pages appear by the courtesy of their owners: Kevin Grace, 13; Wanda Mosbacker, 32 and 33; William Watkins, 41; the Finch family, 47 (William Finch); Radio Club W8YX, 47 (QSL card); Edward Hake Phillips, 54; George Rieveschl, Jr., 55; The Oesper Collection at UC, 65 (Saul Arenson); Ralph Spitzen, 76; Robert Delcamp, 79 (portrait, also on page 107) and 86; James McDonough, 85; Henry Brown, 91; Louis Laushey, 93 (E-Week); Tau Beta Pi, 93 (Bent monument); Lisa Souders, 97 (Sander Hall series); Michael Graves Associates, 102 (building model and plans); Purdue University Archives, 106 (Harry Thomas Cory).

The photographs on the following pages are reprinted with permission from their owners: Golden Gate Bridge Archives, 44; *The Cincinnati Enquirer*, 45.

The photographs in Chapter 6 are by Wendell Day, with the following exceptions: The portraits of Norman Baker and Joseph Steger, on page 90, and the photograph on page 96 are by Don Reimer. The photograph of Dean Papadakis, and Michael Graves on page 101, is by Lisa Souders.

Index

(Boldface figures denote photo or illustration)

H

I

J

K

L

M